29.95

# Student Assessment in Calculus

*A Report of the NSF Working Group on Assessment in Calculus*

©1997 by
The Mathematical Association of America (Incorporated)
Library of Congress Catalog Card Number 97-70507

ISBN 0-88385-152-0

Printed in the United States of America

Current printing (last digit):
10 9 8 7 6 5 4 3 2 1

# Student Assessment in Calculus

*A Report of the NSF Working Group on Assessment in Calculus*

*Alan H. Schoenfeld, University of California, Berkeley, Chair*
*Edward L. Dubinsky, Georgia State University*
*Andrew M. Gleason, Harvard University*
*Del Harnisch, University of Illinois*
*James Kaput, University of Massachusetts, Dartmouth*
*Edward Kifer, University of Kentucky*
*Lawrence C. Moore, Duke University*
*Rogers Newman, Southern University*
*Jane Swafford, Illinois State University*

*Published and Distributed by*
THE MATHEMATICAL ASSOCIATION OF AMERICA

# MAA Notes Series

The MAA Notes Series, started in 1982, addresses a broad range of topics and themes of interest to all who are involved with undergraduate mathematics. The volumes in this series are readable, informative, and useful, and help the mathematical community keep up with developments of importance to mathematics.

# MAA Notes

These volumes can be ordered from:
MAA Service Center
P.O. Box 91112
Washington, DC 20090-1112
800-331-1MAA      FAX: 301-206-9789

*Initiated in 1988, the NSF Calculus Program has had as its goal the reform of the teaching and learning of calculus on a national level. Managed by the Division of Undergraduate Education, the program has had strong support from other Divisions in the Directorate of Education and Human Resources, as well as the Directorate of Mathematical and Physical Sciences through the Division of Mathematical Sciences. The national calculus reform movement is an excellent example of how NSF serves as a catalyst for educational reform on a very large scale.*

*The extent to which faculty and institutions have adopted calculus reform is an important indicator of impact. However, careful assessment of student learning is more critical in determining the ultimate impact and value-added of education reform. This report presents thoughtful perspectives on methods useful for assessing student learning and performance. In particular, the report provides valuable insights into the development of a structure through which student learning can be assessed, illustrates approaches through examples of student work, and identifies areas requiring further research in teaching and learning.*

*The Working Group, under the excellent leadership of Dr. Alan Schoenfeld, is to be commended for their contributions to this volume. Also, the final report was prepared for publication by Carol Quinlan of Compuware Corporation working under contract in the Division of Research, Evaluation and Communication.*

*Director,*
*Research, Evaluation and Communication*

*Director,*
*Division of Undergraduate Education*

# Foreword

The 1990s have witnessed a slow but perceptible increase in the pace of reform in undergraduate mathematics education. According to a recent survey of the Mathematical Association of America (MAA), two-thirds of the more than 1,000 responding postsecondary institutions reported that in the spring of 1994 either modest or major changes in calculus instruction were underway on their campuses. Consequently, it was estimated, nearly 150,000 students (one-third of all calculus enrollments) were involved in some form of reform. The changes ranged from modest efforts such as offering experimental sections with an innovative textbook to revisions of entire courses that entailed extensive use of graphing calculators or computers, open-ended projects, and increased emphasis on modeling and applications (Leitzel and Tucker, *Assessing Calculus Reform Efforts*, MAA Report, 1995, page 5).

The principal catalyzing force in promoting reform has been the National Science Foundation (NSF) calculus program that has been in place for nearly 10 years. And as MAA President Ken Ross has noted (*Focus*, June 1995, pages 3–4), changes in both content and instructional strategies are now spilling over into pre- and post-calculus courses, as well.

For the past several years, NSF has been concerned with finding ways to assess the effects of its curriculum reform initiatives. In the summer of 1992, a group of leaders in calculus reform, including several principal investigators for NSF projects, took part in a meeting on how best to proceed in such an assessment initiative. At the meeting it was quickly evident that a needed first step was to both prepare a conceptual framework for calculus assessment and to provide sample exercises to give meaning to what was being proposed. A working group was formed under the leadership of Professor Alan Schoenfeld of the University of California at Berkeley. This report is the product of that group's efforts.

One fundamental tenet of curricular reform is that in order for significant change to take place, it must be accompanied by substantial modifications in how to measure the planned educational outcomes. A mathematics test is, after all, a statement of what is valued in mathematics education. A set of well crafted test items says more about our educational philosophy and goals than pages of text and data tables can ever communicate.

This report has both theoretical and practical importance in collegiate mathematics education. For researchers, the framework that appears here maps out several keys areas in teaching, learning, and educational measurement that need sustained investigation. For instructors, the document provides examples to illustrate how the extent and nature of student learning can effectively be assessed.

Alan Schoenfeld and his colleagues are enthusiastically thanked, and commended, for their creative and diligent efforts to improve the teaching and learning of undergraduate mathematics.


Kenneth J. Travers

Professor of Mathematics Education, University of Illinois at Urbana-Champaign (UIUC). Director of the Division for Research, Evaluation and Dissemination, 1990–1993.

## Table of Contents

# 1. Introduction

To understand recent changes in collegiate mathematics and the need for new approaches to assessing students' knowledge of mathematics, one needs to understand the character of the curriculum change that has taken place in the United States over the past 15 to 20 years. Hence we begin this report with some broad context-setting remarks regarding mathematics instruction in general. We then focus on the calculus and related courses.

The past two decades have encompassed tremendous progress in mathematics education – progress toward an understanding of the nature of mathematical thinking and learning, toward setting appropriate goals for mathematics instruction at all levels, and toward course development aimed at achieving those goals. On the research front, one need only look at the recently published *Handbook of Research on Mathematics Teaching and Learning* (Grouws, 1992) to see how the field has blossomed. On the development front, insights from research have played out in new goals for instruction and student learning. Thus, for example, in the National Council of Teachers of Mathematics' (NCTM) (1989) *Curriculum and Evaluation Standards for School Mathematics,* the first four curriculum standards at all grade levels are concerned with mathematics as problem solving, mathematics as communication, mathematics as reasoning, and mathematical connections. Of course, reasoning, solving problems, making connections, and being able to communicate mathematically have always been central parts of mathematics. However, until recently they received little overt attention because instruction focused almost entirely on content – on the mastery of mathematical facts and procedures – with little effort to convey a deeper sense of the mathematical enterprise or its place in an invigorated educational experience for all students. The negative effects of such a narrow focus are now widely recognized. Efforts at developing richer, more engaging, and ultimately more mathematical curricula are under way from kindergarten through college.

Efforts at the K-12 level have been spearheaded by *Everybody Counts* (National Research Council, 1989), *Reshaping School Mathematics* (Mathematical Sciences Education Board, 1990), and the two volumes of the NCTM *Standards* (1989, 1991). The *Professional Standards for Teaching Mathematics* (NCTM, 1991), in particular, make it clear that the very character of the classroom environment must be reconceived – that "teaching as telling" is inherently limited, and that it is incumbent upon mathematics instructors to find ways to have their students engage more directly in and with mathematics. Equally important is the role of assessment as a means of helping us to understand what our students are understanding – that is, whether they are making progress toward the goals that have been set for them. Thus Dr. Frank Press, who was then President of the National Academy of Sciences, writes the following in the preface to *Measuring Up* (Mathematical Sciences Education Board, 1993):

> As the United States moves resolutely towards standards-based education, we must learn anew how to measure quality. . . . Assessment demonstrates the real meaning of "standards." We can see in the tasks children are expected to perform just what they must learn to meet our national goals. (p. iii)

If, after all, "what you test is what you get" (see, e.g., Madaus et al., 1992), it is essential to focus on essential aspects of mathematical thinking and mathematical performance in assessment. *Measuring Up* itself, the National Science Foundation (NSF) assessment initiatives in school mathematics and science, and projects such as the New Standards Project, all reflect efforts to develop a better understanding of student learning and performance. Whether you consider a "reform" course, a traditional one, or something in-between, the issue is the same: teachers need to be able to assess the broad spectrum of understandings that their students need to develop.

There are precisely analogous issues in collegiate mathematics. Spurred on by dissatisfaction with current practices and by the opportunities afforded by new technologies, the mathematical community has turned with renewed attention to educational reform – and reform in calculus has been at the forefront of curricular change in general. Through the 1980s there had been increasing dissatisfaction with the "standard" calculus course. There was also an increasingly serious challenge to the idea of calculus being *the* introductory college mathematics course, a challenge spearheaded by the proliferation of courses in discrete mathematics. It was in this historical context that a 1986 conference on the state of calculus instruction produced a clear call for change. That conference resulted in the volume *Toward a Lean and Lively Calculus* (Douglas, 1987). Its message was echoed loudly in *Calculus for a New Century* (Steen, 1987). NSF responded to the calls for reform by sponsoring a major calculus initiative – one that produced a broad spectrum of efforts exploring alternative ways to conceive and teach calculus.

Those efforts were, of necessity, of many different types and with many different emphases. All focused on having students obtain a deeper conceptual understanding of a selected group of calculus concepts, and, to some significant degree, de-emphasized the mechanical, computational aspects of calculus. Some emphasized connections; some emphasized writing and/or projects; some made heavy use of technology (including "laboratory-based" approaches to the course), while others did not. This diversity was entirely appropriate, given the state of the art and the fact that, despite the allegedly monolithic structure of *the* calculus course from the 1950s through the mid-1980s, versions of the course aimed at different populations (e.g., for business or engineering majors) had markedly different emphases in content. With the recognition that (a) multiple constituencies might require different kinds of courses, and (b) there was no single "correct" alternative with regard to either content or pedagogy, the mathematical community explored widely and productively. One can see the results of NSF funding, for example, in many of the publications on calculus. These include *Priming The Calculus Pump: Innovations And Resources* (Tucker, 1990); MAA notes # 17, *Student Research Projects in Calculus* (Cohen et al., 1992); *The Laboratory Approach To Teaching Calculus* (Leinbach et al., 1991); *Symbolic Computation in Undergraduate Mathematics Education* (Karian 1992); *Assessing Calculus Reform Efforts* (Tucker and Leitzel 1995); and numerous articles in *UME TRENDS*.

Following some years of development, it was appropriate to take stock – not so much for evaluation (deciding if efforts were successes or failures, in essence assigning a grade to the reform effort), but rather for assessment: to see what had been learned, to determine if what had been learned could be broadly applied, and to determine what steps might next be taken. Toward those goals, NSF convened a Workshop on Assessment in Calculus Reform Efforts in July 1992. That workshop brought together educators and mathematicians who had been involved in the calculus reform movement and in understanding mathematics thinking and learning to discuss the state of the art. What were the reformers' goals? What had they tried, for what reasons? What had they learned? And what additional information was needed in order to make continued progress?

Two things became clear at that meeting. First, the issue was not limited to reform calculus: the issue was student learning in *any* calculus courses or courses related to them. The field needed better ways to examine student understanding, and all students and faculty would profit from having access to such tools. Second, there was a clear need to develop a broad and coherent way of examining student understanding in calculus – to clarify goals of instruction and to develop and refine ways of examining student behavior to help the mathematical community understand whether and how new forms of instruction are helping students attain those goals. To that end, NSF convened the Working Group on Assessment in Calculus, which met in November 1992 and July 1993 to develop an approach to student assessment in calculus. This report is an outgrowth of those meetings, drawing upon the work of the calculus projects and produced in consultation with their staffs.

The main goal of this report, then, is to outline the state of the art of calculus assessment and to indicate directions for explorations in assessment that will enable us to gain a deeper understanding of student learning and to improve student learning in calculus instruction. The report does so in the context of the two major changes related to calculus instruction that have taken place over the past decade: a revised set of goals for instruction, and a much expanded research base regarding the nature of students' understandings of mathematical concepts. It is grounded in the principle that assessment requires an understanding of what it means to understand — and the recognition that the progress of the past decade provides a good starting point for calculus assessment, but that a much deeper understanding of student thinking and learning in calculus is called for.

In sum, the focus of this report is on the assessment of student learning and the instruction that fosters it. As such, it can be used as a vehicle for thinking about individual student work; it can be used for thinking about goals for instruction, and about how well a particular course or set of courses is meeting those goals. This report is not directed at large-scale policy issues (e.g., was the calculus reform initiative a success? are new courses cost effective?), although an understanding of the topics discussed here will help contribute to a thoughtful discussion of those issues.

# 2. User's Guide to this Report

Our expectation is that this document will be of interest to at least three major groups:

- Calculus instructors and others who are concerned with student learning and understanding in calculus and related courses;

- Scholars in all fields who are concerned with assessment and the development of broad frameworks for assessment; and

- Researchers in mathematical thinking, learning, and assessment, with a focus on calculus in particular.

Here, in brief, is what the report offers each group.

Calculus instructors and all others with a direct interest in calculus instruction will find in the second part of Section 3 a clear delineation of the major goals of the calculus reform projects. Understanding those goals will help those who want to move in the direction of the new courses, those who want to compare such courses (which have different means and ends than traditional calculus instruction) with the traditional courses, and those who want to reflect on the goals of their own courses (whether reform or not) and to examine how well their students are attaining them. In Section 5 the latter group of readers will find a description of major issues one faces when trying to assess a broad range of student understandings in calculus. In Section 7 they will find a collection of assessment items that exhibit assessment techniques and ways of thinking about student performance in calculus. Some student work is also included in Section 7. Reading about and working through the examples – better yet, trying the relevant ones with students and reflecting on the results – will, we hope, provide instructors of calculus and related courses with some practical tools and a useful window into student learning.

Those with a general interest in assessment – or perhaps in assessment in other mathematical or scientific content areas – will find Sections 3.1, 4, and 6 of particular interest. Section 3.1 offers a general framework for assessment that is applied in this document to issues of calculus but is also used for all of K-12 mathematics and has the potential to be applicable in any assessment context. Section 4 provides a brief description of what is known about assessment techniques in general. Section 6 provides a delineation of calculus-related assessment research issues that need to be addressed. Many of these issues transcend the calculus – as, in fact, do the lessons to be learned from the calculus assessment items discussed in Section 7.

Researchers in mathematical thinking, learning, and assessment should find all of the sections mentioned above to be of interest. The goals for calculus begin to specify arenas of focal interest for research into student cognition and into assessment. The framework and the sample assessment items should help clarify assessment issues and techniques. Finally, the delineation of needed research – the focus of the research and development agenda described in Section 6 – highlights a range of studies that should be attractive to researchers and stimulate work in the area. An enhanced understanding of student learning and ways to assess it will serve us all well.

# 3. Framework for Assessment:
## Goals for Instruction and Student Assessment in Calculus

This section of the report offers a general framework for assessment, and then an application of that framework to calculus assessment in particular. In reading the framework, it is worth noting a chicken-and-egg issue of genesis: the need for new assessment methods was occasioned by the emergence of a much broader set of goals for mathematics instruction in general. Hence, goals related to process, attitude, an ability to see and make connections, writing mathematically, and so on, played a major role in shaping conceptions of what ought to be assessed.

## 3.1 A Framework for Assessment of Student Understanding in Mathematics

The dimensions of assessment discussed here are adapted from "*A Framework for Balance*" (Schoenfeld, 1992). The following dimensions should be considered when developing a task or collection of tasks — an "assessment package" — for any particular mathematics program.

*A. The overarching philosophical and pedagogical goals of the program.*

An assessment task should reflect the philosophy and goals of the program. In some programs these goals are explicit, and in others they must be inferred from the problems and activities developed for students.

*B. The content on which instruction focuses.*

Assessment tasks should reflect the following:

- Major concepts students are expected to learn;

- The kinds of sense-making in which the students are expected to engage (e.g., quantitative, spatial, symbolic, relational, probabilistic, logical);

- The major procedures and techniques students are expected to know;

- The kinds of representations students are expected to be able to employ (sketches, tables, graphs, matrices, etc.); and

- The kinds of connections, within and outside mathematics, that students are expected to be able to make.

*C. The kinds of thinking processes students are expected to learn to use and to demonstrate (some or all of the following list).*

- Exploring, experimenting, investigating;

- Formulating, conjecturing, hypothesizing;

- Analyzing, interpreting;

- Evaluating, comparing;

- Planning, organizing;

- Designing, making;

- Justifying, proving;

- Generalizing; and

- Reflecting, explaining, summarizing.

*D. The kinds of products students are expected to produce to demonstrate their competencies (some or all of the following list).*

- Mathematical models;

- Plans or designs;

- Pure or applied investigations and reports;

- Decisions and justifications for them;

- Explanations of concepts;

- Routine problem solutions;

- Exhibitions of technique; and

- Proofs and mathematical justifications.

*E. The kinds of situations students are expected to be able to deal with, e.g.:*

- Pure mathematics problems;

- Illustrative applications (such as standard max-min problems, or conventional applications of linear programming); and

- Complex real-world situations that must be modeled and "mathematized."

*F. Issues of diversity and access; differential performance.*

Assessments should be free of bias and should not favor one group over another in terms of interest or context. New instruction and assessments tend to use more writing and open-ended problems than their predecessors, for example. This raises questions such as the following:

- Are some students empowered, some disenfranchised by these changes? (Note, of course, that some students may well have been advantaged or disadvantaged by the standard assessment techniques; also that background plays a role in that, for example, students who have strong physics backgrounds are at an advantage in calculus courses that emphasize applications to physics.)

- Do some groups of students do better on some types of assessments than others?

- How much is task selection a factor in performance? (That is, do different tasks that ostensibly assess the same notions tend to enfranchise some students and disenfranchise others?)

- How is the fairness of the assessment affected by the scoring procedures, topic selection, and potential for bias in scoring complex responses?

*G. The circumstances under which students produce the work by which they are assessed, e.g.:*

- What the time constraints are;

- Whether work is individual or collaborative, and if there is the opportunity for feedback and correction;

- What resources (technology, information) are available to students; and

- Who generated the task (the students or the instructor).

*H. Issues of perceived value.*

Is the task/the course interesting, engaging, and intellectually gratifying? Do retention rates or performance in other courses reflect well on the students' experiences?

We note that this kind of framework can be used in two ways. First, the framework can be used to examine any particular assessment task. One can look at a task and ask how it stacks up on any of these dimensions. For example, what content or skills does it ask the student to demonstrate? What connections does it ask the student to make? What kinds of things is the student asked to produce?[1] Going through this exercise (as we do, in Section 7) provides a way of examining the richness of any task (or exposing the lack thereof) and serves as a reminder of ways that the task might be improved. Second, the framework can be used to examine a collection of tasks – a proposed "assessment package." There, one can look broadly at the coverage of the collection of tasks and discern what philosophy or pedagogical approach or assumptions they represent (often implicitly). What is the content focus, and is it fair and balanced? How broad is the range of processes students are expected to use, the range of products they are expected to generate, and the set of tools to which they have access? Does it provide opportunities for students to demonstrate their work in a variety of circumstances?[2]

---

[1] Of course, one should not assume that what we see in a task is what students will see in it. The full examination of an assessment task requires field testing to determine what students do when they are asked to work the task.

[2] We note that the framework is intended to be used in a productive way, to help those who are developing or using assessments to see if they are balanced and if the process and content coverage in those assessments reflect the goals of instruction. However, the framework can also be used in an evaluative sense. For example, applying it to standard calculus final exams (especially multiple-choice exams) is likely to reveal just how little mathematical behavior of interest is actually being assessed by the exams.

With these comments serving as a preface, we turn our attention to the specifics of assessment in calculus.

## 3.2 Framework Applied to Calculus, with a Focus on NSF-Supported Calculus Projects

A major focus of the July 1992 Assessment Workshop was a discussion of the goals of the various NSF-supported calculus projects. Subsequent to the Working Group's November 1992 meeting, group members undertook the task of distilling the goals of the various projects, in consultation with project staff. That distillation is as follows. Broadly speaking, the projects were in agreement about the overarching set of goals for instruction listed below. However, as one would expect, given the significant variations from project to project, there was considerable disagreement about the relative importance of the goals and about the best ways to work toward them.

*A. Overarching philosophical and pedagogical goals.*

In large measure the overarching goals of the calculus projects are reflected in the goals for collegiate mathematics instruction summarized in the MAA's *Source Book for College Mathematics Teaching* (Schoenfeld, 1990, p.2).

### Goals

- Mathematics instruction should provide students with a sense of the discipline – a sense of its scope, power, uses, and history. It should give them a sense of what mathematics is and how it is done, at a level appropriate for the students to experience and understand. As a result of their instructional experiences, students should learn to value mathematics and to feel confident in their ability to do mathematics.

- Mathematics instruction should develop students' understanding of important concepts in the appropriate core content. Instruction should be aimed at conceptual understanding rather than at mere mechanical skills, and at developing in students the ability to apply the subject matter they have studied with flexibility and resourcefulness.

- Mathematics instruction should provide students the opportunity to explore a broad range of problems and problem situations, ranging from exercises to open-ended problems and exploratory situations. It should provide students with a broad range of approaches and techniques (ranging from the straightforward application of the appropriate algorithmic methods to the use of approximation methods, various modeling techniques, and the use of heuristic problem solving strategies) for dealing with such problems.

- Mathematics instruction should help students to develop what might be called a "mathematical point of view"—a predilection to analyze and understand, to perceive structure and structural relationships, to see how things fit together. (Note that those connections may be either pure or applied.) It should help students develop their analytical skills and the ability to reason in extended chains of argument.

- Mathematics instruction should help students to develop precision in both written and oral presentation. It should help students learn to present their analyses in clear and coherent arguments reflecting the mathematical style and sophistication appropriate to their mathematical levels. Students should learn to communicate with us and with each other, using the language of mathematics.

- Mathematics instruction should help students develop the ability to read and use text and other mathematical materials. It should prepare students to become, as much as possible, independent learners, interpreters, and users of mathematics.

Many project directors feel that changes in pedagogical style are particularly important. They emphasize that students should be active learners and should learn to think autonomously. They believe that students should work together in small groups, because small group work tends to replace competition with cooperation, promotes conversations about mathematics, and provides a chance for students of different strengths and learning styles to contribute to the solution of problems. Moreover, they argue that students should acquire the habit and skill of working in teams because that will often be expected in later life. Correspondingly, there is a de-emphasis and devaluing of lecturing in the classical style. Many say it should be reduced; some say it should be abolished.

There is general agreement that students should work on some ill-defined and open-ended problems, to learn that often one must construct and test assumptions not explicitly stated in a problem in order to find a solution. The expectation is that students should realize that significant problems usually take more than a few minutes to solve, and that they should learn to accept the frustration that accompanies such work (and the corresponding gratification when a difficult problem yields to their efforts).

Broadly speaking, project directors believe that today's students will almost always have computers and calculators available, so they should learn to use such technological devices appropriately. Some project directors noted that computers also can and should be used to enable (or encourage) students to construct their own mathematical understandings. In addition, it may well be the case that students will (appropriately) develop a *different* sense of the domain as a result of the accessibility of technology — a sense that corresponds closely to some contemporary technology-based uses of calculus.

A common goal is that students should enjoy learning, doing, and applying mathematics, and thus be encouraged to study more mathematics. Project directors want their students to be able to use what they have learned in subsequent courses, both in and out of the mathematics department; they also want their students to retain some understanding and knowledge for a long period of time and to be able to learn more mathematics on their own.

*B. Content.*

In broad terms, there is a consensus that most calculus courses as currently taught are crammed far too full with material, with the result that topics are given too superficial a treatment. As a result, there is the wish that the mathematical community guard against teaching topics because "we've always taught them," and instead converge on a small number of topics to be learned well. (This is, we note, an echo of the themes sounded in the *Lean and Lively Calculus* (Douglas, 1987)). Correspondingly, there is a consensus that any focus on techniques should be kept subordinate to having students develop an overall view of the subject.

The list of content goals delineated below was compiled with the expectation that students would be taking a sequence of courses extending over three or four semesters. Goals preceded by an asterisk (*) are those on which there is general agreement.

- The following content areas should be *emphasized*:

  * introductory concepts, including subsets of $\Re^2$ and $\Re^3$ and simple transformations of these spaces; vectors, etc.

  * functions and representations of functions, including numerical, graphical, symbolic, and algorithmic representations.

  * limits and continuity from an intuitive and constructive point of view.

  * the derivative of a function as a function that gives the relative rate of change or the slope of a tangent.

  * parametric representation of curves.

  * the Mean Value Theorem in an intuitively clear form, such as

    "If f' > 0 on an interval, then f is increasing on that interval."

  * the integral conceived of as an accumulation of many small quantities that provides a means of calculating a total change from a rate of change, area, volume, and other geometrical and physical quantities.

  * both parts of the Fundamental Theorem of Calculus, with an emphasis on the conceptual distinction between indefinite and definite integrals.

  * integration by parts and straightforward substitutions; the use of integral tables for more complex integrals.

* numerical integration and numerical differentiation.

* conventional third semester topics with an emphasis on surfaces and vector calculus.

  improper integrals.

  ordinary differential equations with emphasis on initial value problems and Euler's Method for numerical solutions.

  systems of differential equations.

  other numerical procedures, e.g., Newton's Method, approximation of functions by Taylor's or Fourier series with the resulting introduction to series. (There was a consensus that geometric series should be singled out and given a more prominent role.)

  applications of calculus to probability and statistics.

* There are arguments that the following content areas should be *de-emphasized*:

  epsilon-delta arguments about limits, continuity, etc.

  the Mean Value Theorem in its classical form.

  tests for convergence of series beyond comparison with a geometric series.

  techniques of integration (other than simple substitution and integration by parts).

## C. Thinking processes.

All of the projects agree that there should be an emphasis on teaching skills of an order higher than solving template problems. Project directors felt that students should learn

* to deal with situations that are unfamiliar, but mathematically related to topics they have studied;

* to experiment with examples to discover relationships, and to make and test conjectures;

* to express problem situations in terms of functions;

* to construct and analyze models—discrete and continuous, empirical and theoretical — and understand how to use them;

* to make successive approximations to the solution of a problem;

* to use top-down, divide-and-conquer design to solve complex problems;

* to combine intuition, generalization, and logical arguments to find solutions and to explain why they are correct (here the emphasis was on the explanation rather than on formal logic);

* to form mental images or representations of mathematical processes such as forming the inverse of a function or taking its derivative; and

* to visualize mathematical objects. Students should regard curves and surfaces as important and useful objects in their own right, not as the end products of tiresome exercises.

The expectation is that students should learn to reflect on problems both before and after attempting to solve them. More broadly, the expectation is that they should learn that mathematical understanding is the result of working on problems, reading mathematics, and talking about mathematics. Various projects (and the assessments they employ) will differ in the degree to which they emphasize the thinking processes highlighted here and in the general list described above. We note that this is natural: the purpose of the framework is to provide a list of important candidates, thereby enabling instructors to see which things they are focusing on and which they are not.

## D. Student products.

As one would expect, the calculus projects differed in their expectations of the products they expect students to generate (for example, some placed a heavy emphasis on mathematical models, or on reports of exploratory activities; others did not). There was a consensus that in addition to being able to formulate and solve a problem using the concepts and techniques of calculus, students should be able to describe the process of solving it and to communicate the solution both orally and in writing. Some projects expect such a report not only to tell how a solution was obtained but also to include a description of ideas that were tried but that proved unsuccessful.

The projects share an increased emphasis on oral and written communication. However, there is significant variation across calculus courses in the size of the pieces of work students are expected to produce and in the amount of writing expected from students. In some projects, lucid explanations of standard (or slightly non-standard) homework problems, reflecting perhaps an hour or two worth of work, are the norm. In others, students are expected to produce extended laboratory reports or reports of investigations that last over a period of weeks.

*E. The kinds of situations with which students are expected to deal.*

Some project directors expect their students to learn that calculus applies in many contexts outside of the hard sciences, including management, life, and social sciences. Merely telling this to students is not likely to be effective, so there is an obligation to have students work on problems from these domains. (Note that working such problems could be in the form of routine applications, or it could be in the form of "mathematizing" complex situations from those domains.) Likewise, there is some sentiment that students should realize that both approximate and exact solutions have value (and a concomitant obligation to provide relevant experiences for students), and there is the wish that students understand the distinctions between models and the situations they represent.

*F. Issues of diversity and access.*

There is uniform agreement that calculus, like all of mathematics, should be "a pump, not a filter"; that the study of calculus should become more attractive and more accessible to women and minorities; and that all students should be given more opportunities to succeed. The issues for assessment are both department- and program-specific (e.g., does a particular department or program do well with regard to access and equity goals; what are the success rates of women and minorities?) and program-general (e.g., are there features of programs, in general, that tend to enhance the success of women and minorities?).

*G. The circumstances under which students produce the work by which they are assessed.*

In general, assessment should be consistent with the goals of instruction and with the students' experience. For example, one only discovers whether students can do collaborative work, or use technology, only if assessment provides them an opportunity to do so. Moreover, as discussed below, what is assessed signals to students what is valued — that is, what really counts.

# 4. What We Know About Assessment in General

In brief, much less is known than we would like, but the picture is changing rapidly. There are now the beginnings of major changes in K-12 assessment – largely grounded in an increased understanding of the processes of mathematical thinking and learning – which are expected to continue over the next few years. Research on learning (both in general and in particular with regard to the fundamental concepts of calculus) and work in K-12 assessment point to new directions for calculus assessment as well. It is hardly the case, however, that all the relevant issues have been identified, or that the appropriate methods are known. New methods must be developed in order to bring assessment more in line with evolving instructional goals, and additional research will be needed to undergird the assessment efforts. The ties between research on learning and research and development on assessment should be close indeed.

The recognition that the mathematical community needs fundamental changes in assessment methods at all grade levels derives from an increased understanding of what it means to think mathematically – the kind of research-based understanding that produced the refined goals for instruction reflected in the NCTM *Standards* and the MAA's Goals for Collegiate Mathematics Instruction (Schoenfeld, 1990, p. 2), which were quoted on page 9 of this report.

Until recently, the mathematics that a student was deemed to understand was defined in terms of a "content inventory." To assess the student's knowledge, one checked for the student's mastery of particular bits of content. That made assessment simple (though wrong-headed). For K-12 mathematics, one looked at a list delineating the aspects of arithmetic, algebra, and geometry that students were expected to learn; for the Graduate Record Examination (GRE) of advanced mathematics, specifications included a certain percentage of problems from calculus, linear algebra, analysis, and so on. Then, an appropriate sampling of short items (each requiring only a few minutes for solution) corresponding to the list of topics was assumed to provide an inventory of what the student knows. In addition –

and a huge amount of research time went into the issue – such content sampling could be done in ways that met the psychological definitions of "reliability" and "internal validity." That is, it was possible to construct examinations with the property that when students took the same test twice, or took alternate versions of a test, their scores were essentially the same. Hence alternate versions of the test could be said to measure the same thing (so that, for example, a score of 740 on this year's GRE or Scholastic Aptitude Test (SAT) is taken to mean the same as the same score on last year's exam, or even on the exam given a decade ago). Moreover, difficulty could be adjusted by raising or lowering the technical skill level required to solve particular problems (e.g., by increasing the number of nested functions in a chain rule differentiation).

However, as the community has come to recognize the complexity of mathematical thinking and learning over the past few decades, it has also come to recognize the gross inadequacy of the "content inventory" approach. This approach is founded on assumptions about the nature of knowledge that are no longer justified. It addresses only category B of the framework described in Section 3.2 and thus ignores other fundamental aspects of mathematical performance. Moreover, because of the ways in which assessment drives curricula (and shapes students' understanding of what is important), such narrow content-oriented assessments distort and derail the educational process. For that reason, a large number of alternative assessment devices are being developed. Those devices, in large measure, have the advantage that they are more consistent with new goals for instruction. They have the disadvantage that as new ground is being broken, the developers have had to forsake the old narrowly defined standards of reliability and validity. In large measure, such notions will have to be reinvented. (We shall have more to say about this issue in our discussion of technical considerations, following the discussion of various types of assessment methods.)

In what follows we describe a broad range of assessment techniques that apply to all kinds of instruction –

precalculus as well as calculus or upper division courses, traditional as well as reform courses. The methods described range from paper-and-pencil items to complete student portfolios that document student achievement over a year's time. In general, they reflect a move away from conventional practices, where the primary use of testing was to obtain scores for the purpose of assigning grades, to the use of tasks that are meaningful and instructionally worthwhile in their own terms — tasks more consistent with the notion of "authentic assessments" as characterized by Wiggins (1989). Three commonly occurring features of such assessments are the production of discourse, objects, or performances; flexible use of time and other resources; and collaboration with others.

## Category 1: Various Pencil-and-Paper Assessment Tasks

Depending on class size and the resources that are available, calculus instructors have tended to rely on a mixture of multiple-choice, short-answer, and open-ended or essay questions. That mixture has ranged from 0 to 100 percent in any of the categories. One can find departments where all examinations are multiple-choice (and machine-graded), others where short answers form the majority or the totality of an exam; and yet others where the exams are all essay questions, which receive detailed readings for partial credit.

### Multiple-choice and Short-answer items

These two kinds of items tend to be objective (though limited in scope), efficient, and reliable. In large lecture classes they are expedient, but an intellectual price must be paid for this expediency: the use of such assessment tends to undermine the goals for instruction stressed in this report. A response to a multiple-choice or short-answer item often requires no more than a computational procedure or the statement of a definition or fact. Indeed, a study of college calculus syllabi and final exams from 1945 to 1990 found that 65 to 75 percent of the first semester calculus exam items could be categorized as symbol manipulations and calculations requiring little deep thought. In the second and third semester calculus exams, the percentages were even higher, ranging from 75 to 95 percent (Dubinsky and Ralston, 1992).

Items like those typically found on multiple choice and short-answer assessments often convey the idea that mathematics is made up of unrelated bits and pieces and that learning mathematics is memorizing rules and procedures or acquiring a bag of tricks. These items rarely assess students' ability to solve problems, synthesize ideas, create new knowledge, or communicate observations. In emphasizing one correct answer, traditional items do not allow for complex answers or for multiple ways of looking at a problem. Further, timed tests, the most widely used in college classrooms, tend to place more value on thinking quickly than on thinking deeply. They tend to discourage persistence and to promote impulsiveness. In the real world, however, speed is often less important to success than reasoning, analysis of a situation, and perseverance.

Some researchers are pursuing alternative, richer ways of exploiting multiple-choice formats. For example, what is called a "power item" consists of a sequence of increasingly difficult questions related to a single problem. According to their advocates (see, e.g., Pandey, 1990) such items have a number of desirable properties including "face validity" — that is, they seem to be getting at something deeper than isolated skills. Power items, like other multiple-choice forms, are efficient and relatively inexpensive. Other researchers, such as Thornton, are exploring ways to develop multiple-choice tests that provide information similar to what they can obtain from detailed interviews with students.

Whatever their utility, assessments based on multiple-choice are significantly cheaper to administer than any of the known alternatives, so we are not optimistic that multiple-choice will disappear from large-scale assessment in the near future. Nevertheless, there are some cracks in the dam: open- form questions requiring written responses are beginning to replace some closed-form items on large-scale assessments. The College Board announced (Elson, 1990) that in 1994, 20 percent of the mathematics section of the SAT, which is taken by 1.6 million students in the United States annually, would be questions requiring students to produce an answer. The 1992 version of the National Assessment of Educational Progress (NAEP) in the United States also included student-constructed response items, and as many as half of the items on the next NAEP mathematics exam may be of that type. The use of questions that require responses provides some advantage over the use of typical multiple-choice questions: e.g., student reliance on some test-taking skills, like eliminating implausible answers and guessing, is minimized. However, the scope of open-form questions is typically quite limited. The problems still have the property that they admit one "right" answer, and the fact that one sees the student's answer but not the student's work means that one has no real evidence about how the student attained the answer. Thus, such assessments do not facilitate movement toward the kinds of instructional goals discussed above.

## Open-ended items

Open-ended items (essay questions) are those with more than one correct answer or multiple paths to a correct solution. In responding to such items, students are often asked not only to show their work, but also to explain how they got their answer or why they chose the method they did.

There is a wide range of ways in which such items are used, and in the kinds of information one can glean from them. At the more routine end of the spectrum we simply ask to see students' work on standard exercises: say on typical max-min, related rates or volume problems. Their work provides evidence of mastery of basic concepts and some subsidiary skills (e.g., finding an algebraic representation of a situation described in words, or drawing a 2-D sketch of the surface obtained by rotating a curve around an axis). However, if the problems are isomorphs of text examples and homework exercises, they provide little evidence of students' ability to think. At the other end of the spectrum, a 20-minute essay question can provide students with some opportunity to experiment, explore, or document some important abilities such as explaining an important idea. One need not be limited by the routine or the algorithmic in assessment. What follows are two examples of somewhat different character.

The first problem comes at the very beginning (Section 1.1, problem 1) of the Harvard Calculus Project's *Calculus Preliminary Edition*. It appears as a homework exercise (which is, of course, one form of assessment), but might serve just as well as a test or interview item.

> Match the stories with three of the graphs in Figure 1 and write a story for the remaining graph.
>
> (a) I had just left home when I realized I had forgotten my books so I went back to pick them up.

> (b) Things went fine until I had a flat tire.
>
> (c) I started out calmly, but sped up when I realized I was going to be late.

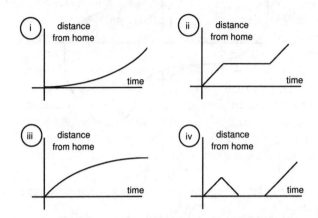

Figure 1. Which story goes with which graph?

The content of this problem is elementary — it comes, after all, at the very beginning of the course — but it signals a difference in intention and style from the conventional approach. First, it lets students know they will be held accountable for a major theme in the course: students are expected to become familiar with different representations of mathematical phenomena (in this case, graphical and verbal) and to be able to interpret them and translate between them. Second, it emphasizes qualitative reasoning in a course that is typically considered to be purely quantitative. Third, the open-ended character of the last part of the problem — "write a story for the remaining graph" — provides students with the opportunity to generate their own stories. There is clearly no single right answer, and the stories students generate are likely to provide some insights into their understandings.

The second problem, which is more advanced, deals with issues of function definition and composition. It is a slightly modified version of a test problem from the C[4]L project at Purdue.

> (a) Define a function **g** whose domain and range are subsets of $\Re$ and whose graph is a subset of the curve that is given in Figure 2.
>
> Explain how your choice defines a function and discuss some other choices that would not work.

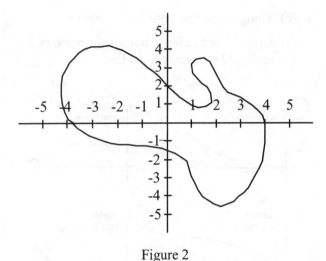

Figure 2

(b) Let f be the real-valued function defined by $f(x) = \sqrt{x^2 - 4}$, over the largest subset of $\Re$ for which that expression is defined.

Find a solution to part (a) of the problem such that the composition **f o g** has the same domain as **g**. Sketch a graph of the composition.

There are many ways to obtain a function **g**, leading to (in some cases) very different examples of **f o g**. Give several such **g**'s and discuss the different possibilities they imply for **f o g.**

(c) Choose a solution to part (a) of the problem with the property that x = 3 is in the domain of **f o g**. Estimate **f o g** (3), and explain how you arrived at your estimate.

This problem, like the one that preceded it, allows the students a fair amount of latitude in determining their answers and justifying them. As a result – especially if the students' written work is followed up with individual or classroom discussions – faculty can learn a great deal about students' understandings (of function definition, of domain and range, of composition, of the meaning of the graph of a function, and of a relation), about what they think is important (e.g., do they try to define **g** for a relatively large domain?[3]), and about the character of their justifications (how do they explain that a subset of

the curve does not define a function, or that a particular function composition is not defined?).

We have chosen these two examples because they are, in some ways, not that different from standard examples. They deal with concepts of central importance and ask students to display competencies that most faculty would agree are reasonable to expect. Yet, in their somewhat open-ended character, they provide opportunities to delve into student understandings, both for purposes of diagnosis and instruction. In general, open-ended items that call for qualitative interpretations, modeling, and other deep mathematical skills can provide a fair amount of information about students' abilities to deal with the nonalgorithmic aspects of mathematics.

<u>Student-constructed tests</u>

Another way to assess students' understanding of a subject is to have them, individually or in groups, develop their own test for a unit or a chapter of study. A test constructed by students identifies for the instructor which content students view as important and the depth at which they perceive their understanding of that content to be. (Such information can be shocking, and informative.) The test construction process itself is a learning process, with students solidifying their understandings or teaching each other.

**Category 2: Performance Assessments of Various Types**

The move to more authentic assessment encourages the use of observations of actual student performance in addition to pencil-and-paper assessment tasks. The kinds of tasks described below all fall in the general category of performance assessments. Much of our experience, and some of the examples given below, are derived from K-12 mathematics.

<u>Performance tasks</u>

When the form of assessment is a performance task, students are presented with an open-ended task and observed as they go about dealing with it. Often

---

[3] One student, for example, wrote, in essence: "I could just circle the point (3, g(3)) on the top part of the graph. But that would be cheating." She then went on to define **g** by specifying a maximal subset of the curve that defined a function. Her answer revealed a significant amount about what she understood.

performance tasks involve the use of some physical materials. In one example (Massachusetts Department of Education, 1990), students were provided with a large container of popcorn kernels, several smaller containers graduated in milliliters, a pan balance, and weights. They were asked to estimate the number of kernels in the large container. (They were not allowed to count the entire collection of kernels directly.)

In calculus, a performance task might involve the solution of problems at a computer work station equipped with a variety of computational and graphing utilities. The issues might be quite complex; e.g., students might be asked to model a multivariate real-world phenomenon, explain how the model was designed, run the model, compare its performance with the phenomenon, and revise the model accordingly. (Laboratory exercises from some of the projects are designed along these lines, and many of them could be used as performance assessments.) Note, however, that there are lessons to be learned from much less complex contexts and interactions. A standard problem in a traditional calculus course is to sketch the graph of a function, given its formula. When technology is available, for example, a graphing calculator, sketching a graph may become merely a matter of copying. However, such problems as precisely locating maxima, minima, and points of inflection remain. Moreover, blind reliance on a calculator to draw the graph of a function can lead to seriously erroneous results. Consider the function given by the formula $x^{-12} - 2000000x^{-6}$. This function approaches $+\infty$ as $x$ approaches 0, but if it is graphed on a calculator or computer and there is no pixel whose first coordinate is between 0 and 0.1, the graph will appear to approach $-\infty$ as $x$ approaches 0. This error is not likely to be made by a student who understands the concept of the growth rate of a function, who will see that the $x^{-12}$ term must dominate as $x \to 0$. The student could then change the scale settings to accommodate the missing parts of the graph and finally use calculus to locate the turning points. Another student might apply standard calculus techniques to locate the turning points at x = .1 and -.1; then the true shape of the graph would emerge quickly. As with many problems, asking for a verbal description of the thinking process involved may bring out significant diagnostic information.

## Investigations and projects

Another form of performance assessment is the assignment of investigations or projects. This form of assessment is already used by many calculus instructors.

A wide range of projects and investigations can be undertaken by calculus students. The Consortium for Mathematics and its Applications, Inc. (COMAP) modules, as well as articles in the *College Mathematics Journal* and elsewhere, present a number of interesting topics. Investigations and projects take varying lengths of time to prepare and can be either individual or group endeavors. The usual product is a written report, and where time permits, a class presentation. The volume *Student Research Projects in Calculus* (Cohen et al., 1992) offers an extended collection of such projects, with discussions of how they have been used in the calculus courses at New Mexico State University.

## Observations

Observations need not be separate assessment events but can be incorporated into regular instructional activities. Instructors often formulate reasonable opinions concerning the competencies of their students from informal observations, which formal assessment then legitimizes and quantifies.

Such observations can be done both at the whole class level and the level of individual students. The typical college mathematics class provides few opportunities to observe students' behavior, since students are often passive in the classroom. New instructional arrangements allowing for the active participation of students do, however, provide opportunities to observe students' mathematical performance. In large classes, faculty can periodically ask questions of, and note the responses of, students who are performing at different levels in the course. Such questions are particularly useful if they are keyed to typical student trouble spots with regard to student understanding. They can help the faculty member to calibrate the effectiveness of instruction and to choose among alternate instructional paths.

Individual students can be observed working in small groups, making problem presentations, or posing conjectures or questions in class. The challenge is to structure observations and to develop a systematic procedure for recording them.

Students' attitude toward mathematics is an important goal of mathematics instruction, and a good way to assess progress in this domain is through observation. An inventory form listing desirable dispositions can be kept for each student or for the class. Similar forms can be used as a checklist to observe group work.

Whatever its form, performance assessment can give information about students' understanding not available or only inferred from pencil-and-paper assessment. Such mathematical behaviors as the ability to reason soundly, formulate hypotheses, think flexibly, use tools, use technical terms, collect and organize information, and use estimation can be observed during performance assessment. Desired behaviors such as the ability to communicate mathematical ideas, to work independently, or to work well with others can be observed. Further, student attitudes can be assessed by rating such behaviors as confidence, anxiety, persistence, and enthusiasm while completing the task.

## Interviews

Interviews and conferences with students are a source of rich information about their understandings and feelings about calculus. An interview involves a planned sequence of questions, while a conference implies a more informal discussion with a student. Interviews may contain a sequence of open-ended questions that serve as points of departure for more extended probes of student understanding. Such interviews are time-consuming, but they can help diagnose learning difficulties in a way no other assessment techniques can. While many instructors may not feel that they have time to conduct formal interviews with each student, instructors do have students who come in during office hours for help. These sessions can become interview opportunities by keeping records and by using set procedures such as the five-point error analysis suggested by Newman (1983).

Note that interviews can be used in a variety of ways (as can other forms of assessment), depending on who and what are being assessed. In the most straightforward kinds of interviews, where the goal is to assess student understanding as described in the previous paragraph, one has a good idea of what the student is supposed to understand and probes that understanding in a verbal give-and-take session. But interviews, conferences, and performance tasks can also be used for research purposes. Much of the recent progress in understanding student cognition (e.g., elaborating the complexity of the learning process, delineating misconceptions or "cognitive obstacles," outlining "learning paths" toward the understanding of complex mathematical ideas) has been made by the detailed examination of students as they work on or talk about mathematics. (Sessions are often videotaped, and analyzed *post hoc* at length.) Similarly, such sessions help faculty understand the effectiveness

of the instructional approaches they are trying out—where they succeed, and where they need refinement.

## Category 3: Portfolio Assessment

In addition to student performance, a variety of student products can be assessed. One technique that is becoming increasingly popular is the development of a student mathematics portfolio. While the use of portfolios has been a common practice in art and writing for a long time, they have only recently been used as a method to assess student progress in mathematics. One state, Vermont, is using portfolios in writing and mathematics on a statewide basis to assess K-12 students' growth and understanding over time (Vermont Department of Education, 1992).

What goes into a portfolio should, obviously, depend on the instructional goals of each situation. Typically, a portfolio includes a spectrum of student work – some of which is optional (e.g., "Choose two pieces that you think exemplify your work at its best, and explain why you think they do.") – and some of which is mandatory (e.g., students in a particular course must include one laboratory report, one open-ended exploratory problem, and one report of an extended collaborative project). Typically students are asked to include in the portfolio a cover letter for reviewers that explains why they have chosen the entries they have and what the reviewer should look for in them. Writing such a letter can be a powerful occasion for student reflection on virtually all important course objectives.

If this seems like it is far too much work, it might be worth recalling that a portfolio is, in essence, the kind of documentation that most faculty are asked to put together for their promotion and tenure cases. It allows us to shape the case, to highlight our strengths, and to document the breadth of our achievements (getting grants, providing service of various types, etc.) – while providing some kinds of mandatory evidence (which at different universities might be mandated student evaluations, copies of some or all papers produced during the review period, etc.). Presumably universities could make their promotion and tenure decisions on the basis of less, or less adequate, evidence—but it would most likely make faculty very nervous if they could not make the best case for themselves. Portfolios provide students with similar opportunities.

A portfolio in calculus might include some of the following items:

- Responses to open-ended questions;

- Reports of individual or group projects;

- Work from another subject area that relates to calculus;

- A problem made up by the student;

- Research reports about some facet of calculus;

- Computer-generated examples of student work;

- Student self-assessment reports;

- Journal excerpts;

- A mathematical autobiography;

- Tests or test scores; and

- Instructor's interview notes or observation records.

In contrast to most testing situations, which tend to document what students cannot do, portfolios allow students to document what they can do. This medium enables students to demonstrate the learning and understanding of ideas beyond the knowledge of facts and algorithms. Work in a portfolio can show a student's ability to solve problems, to reason and to communicate mathematically, and to make connections. Portfolios can show students' growth over time and their disposition toward mathematics. Portfolios can be used to assess individual students or to evaluate entire courses or major programs. Collectively, portfolios can reflect the emphases of a calculus program. While more cumbersome and difficult to manage than pencil-and-paper tests, portfolios offer a bigger window through which to view students' mathematical understandings and abilities as well as their growth.

Once again, we note a dual role for such assessments. On the one hand, they provide students with the opportunity to document what they know and what they can do. On the other hand, such extensive documentation of student work provides faculty with a wealth of information about what is effective in their instruction and what might profit from more attention or a different approach.

**Category 4: Self-Assessment**

One of the most constructive and empowering goals of education is to equip students to monitor their own progress. To do this students should be called on to reflect on their own learning and to participate in the assessment process. The capability and willingness to assess one's own progress is the mark of the mature learner. Mathematical power comes with knowing how much one knows and what to do to learn more.

The simplest example of self-assessment is a questionnaire following an activity, a class, or the end of a course that asks students to evaluate the impact of the experience. Students might be asked to identify what mathematics they learned, to relate it to something they had learned before, and to identify any new questions that were raised.

A mathematical journal, offering a daily or periodic record of experiences with mathematics, is another form of self-assessment. In a mathematics journal students might record the most important thing they have learned since their last entry, what they are having trouble with, what they found easy, how they feel about the topic under study, or what they liked most or least about the mathematics class. Journals can be private or shared with the instructor for comments.

With the development and spread of the use of open-ended items, performance tasks, assigned projects, structured observations and self-assessment instruments, there are now a variety of assessment techniques available to the classroom instructors of calculus. Many of these are being used widely in precollege mathematics classrooms or by states in their assessment programs (Stenmark, 1989, 1991). The challenge is to identify what techniques are best for the many goals of calculus and to adapt these alternatives to the content of calculus and the varied realities of college mathematics classrooms.

Technical considerations

Psychologists who specialize in test theory (known as psychometricians) have dealt with the technical properties of (mainly) multiple-choice tests for over half a century. There is an extensive literature on the topic, along with a collection of extraordinarily sophisticated techniques and a general consensus about the desired properties of such tests. From the test-makers' perspective, notions of content coverage are simple – it means taking a sampling of problems from different content domains. And, as long as one focuses only on content mastery, using multiple-choice or similar tests to see if students can get the right answers, then technical issues of reliability and internal validity (dealing with whether the test and alternative versions of it get at the same things and produce consistent scores) are reasonably well defined and understood. Techniques to assess test validity and reliability are well known and available for use by the educational community. These notions are the basis for a set of standards that are and should be applied to traditional tests and measurements.

When it comes to performance assessments, portfolios, and so on, the situation is very different. On the one hand, such assessments are clearly much more aligned with the spirit of reform and seem much more "authentic" and meaningful than traditional assessment measures. On the other hand, the consistency, validity, and fairness of such assessments have yet to be established and codified. Performance assessments presume to measure different things in different ways. They presume different criteria, implicit as well as explicit, for what is good assessment. Hence, new characterizations of desired measurement properties are needed. Significant research is needed to determine which are the salient characteristics of performance assessments and what evidence can be gathered in systematic ways to determine whether a particular assessment has desirable characteristics.

Descriptions of good performance assessments include such terms as fidelity and authenticity—having students do things that reflect the meaningful uses of mathematics. Good tasks are contextualized (they take place in meaningful contexts), engaging, and revealing. Tasks are said to have "face validity"—they appear to tap directly into skills and understandings that we believe are important. Standards for performance are public, with all participants knowing beforehand what is or is not a

good performance. Practicing on tasks is not cheating, but is considered a desirable way to improve performance.

The characteristics of validity and reliability of performance assessments are likely to differ substantially from those of traditional tests. More broadly, the ways in which scores are assigned and interpreted will need careful thought. Writing on that issue in general, Harnisch (in press) raises the following questions, among others: How fair are performance assessments to students of different subgroups? How do we assure that special populations are fairly treated in a performance assessment context? How is fairness of performance assessments influenced by the scoring procedures? By the selection of topics or problem contexts? By potential biases of those doing the scoring? How many meaningful tasks are needed to have confidence in describing student performance and discussing program effectiveness?

There are also, one should note, significant issues of costs and feasibility. "High stakes" performance assessments (that is, those that play significant roles in determining people's futures), to be equitable, must have multiple scorers for complex tasks, so that individual readers' biases can be identified and resolved. But multiple scoring is expensive, and checking to see that different scorers assign the same scores to student work is only a partial solution to the delicate issue of obtaining objective scores for performance assessments. A good deal of research is needed on the development of solid and consistent scoring procedures, the comparability of performances across tasks, and the extent to which there is comparable performance across settings. Implementing such assessments will not be cheap (certainly not as cheap as using multiple-choice tests!), but making such changes is part of the cost of the improvement of education.

# 5. Assessment Issues Faced by the Calculus Projects

Each major calculus project has had to create the means of assessing its more complex and ambitious objectives. Here we provide a general description of strategies developed by various projects to match their new, expanded curricular objectives with new, expanded forms of assessment. (See Section 7 for a collection of assessment tasks.)

In general, most projects have downplayed the role of computational competence, partly by having students use computer algebra systems and partly by simply substituting other activities in place of practice with routine computation. The effect in all cases, but to varying degrees, has been to render computational competence subsidiary to concept development and/or applications. In each of these, students are expected to employ a variety of representation systems across a variety of contexts.

Many innovative programs attack the curricular problem of structuring the objectives, activities, and assessment of their courses by using a layering approach. Learning objectives are organized into several layers, ranging from routine computational skill, to the basic concepts, to relatively straightforward applications of these computational skills and concepts, to higher level modeling and problem-solving abilities. These latter objectives often include the ability to create complex models and to be able to justify and communicate these models using combinations of mathematical representations and writing in English. For example, Project CALC makes the point that the process of writing provides means by which conceptual understanding can be achieved, and that the products of this writing provide means by which it can be diagnosed. Hence, writing is simultaneously a goal, an enabling activity, and an assessment instrument. Indeed, this general pattern of enhanced and self-documenting activities is a hallmark of the newer forms of assessment. Students are assessed on ever more complex activities occurring in ever larger chunks of increasingly independent work.

The assessment contexts, particularly when weighted for importance and impact on final grade, tend to be coextensive with learning contexts. As opposed to the traditional test/quiz situation, where the learning situation and the assessment situation are relatively independent, students are put in the position of learning mathematics by producing a product of some kind (lab report, project report, oral presentation, etc.), which is then assessed. The assessment is thus more intimately linked to learning.

## Layered Activity Structures

As noted above, many projects structure student activity in layered forms that, to a large extent, reflect the layered structure of the learning objectives. More importantly, the weighting of the assessments reflects the revised value attached to these layers. At the lowest levels, computational skill is developed in standard textbook exercises, usually assigned as homework and reviewed in class or discussion sessions. Classes usually are supplemented by weekly computer laboratory activities that involve more extended exploration of particular topics or models. Finally, there are project activities of varying sizes. Some involve only a few days' work, while others act more as term projects. Some involve group work, while others involve individual work. In cases of group work (be it for homework, labs, projects, or tests), grading may be applied to individual contributions or to the whole group.

## Correspondingly Layered Assessment Structures

We see two styles of handling the assessment of routine computational competence. Some embed routine tasks into regular examinations and quizzes but only count these exams and quizzes for a small portion of the course grade. On the other hand, Project CALC factors these tasks out into a separate set of "gateway" competency exams in the second calculus course, which can be repeated, which require a strict 95 percent success level, and which must all be passed before the course can be completed.

The project at Worcester Polytechnic Institute (WPI) likewise factors the more routine skills out into exams, which can be repeated if needed (although repeating tests results in lowering the upper bound on the student's possible exam grade) and which count for only 25 percent of the course grade.

## Assessment as a Signal of What Is Valued

While it is commonly understood by students and faculty alike that tests signal what topics are important in a course, it is less appreciated that the means of assessment also send signals about what is genuinely valued in other dimensions. For example, if computational assistance is not allowed on tests, then it is implied that hand-computation is critically important. Or, if individual work is graded and group work is not graded, then collaboration is being signaled as unimportant or at least not an important objective for the course. If projects are simply checked off rather than seriously graded, then the higher-order intellectual skills that they require are devalued. A similar comment can be made about the way writing, or communication more generally, is treated. Since most projects have a rich collection of objectives, this richness is reflected in their modes of assessment and the ways value is assigned to their assessment activities. This richness can reach impressive levels, as, for example, in the Calculus in Context materials, where respect for qualitative thinking and student ability to deal with the messiness of real problems is given explicit attention in an extended manner across the courses and in the take-home examinations.

## Assessment of Group Work as Enhancement of What Is Valued

Consider, for example, the goal of promoting collaborative work. It is possible to put students in the position of forming a group for a laboratory project, for example, without actually stimulating any meaningful collaboration beyond, perhaps, apportioning a lengthy set of exercises to be done separately or in tandem. On the other hand, it is possible to put students in the position of deliberating a difficult issue, where the deliberation itself plays an important role in learning about that issue. For example, in one lab activity at Worcester Polytechnic Institute (WPI), students, who had dealt with the weaker form of the Mean Value Theorem (assuming a continuous derivative on the closed interval rather than simple differentiability on the open interval), were asked to find a function that failed to have a continuous

derivative. The work of those students gave evidence of their having collaboratively struggled with the question, even though they did not find an example. Similar effects can be had with messy or ambiguous modeling problems that involve important judgments regarding starting point assumptions. As with the case of writing discussed above, multiple objectives are addressed simultaneously: the act of arguing out the assumptions deepens students' understanding of the problem situation while simultaneously giving them the experience of working in a group.

As noted in the previous paragraph, assessment methods signal to students the seriousness with which their instructors take group work. Grading schemes range from having individual write-ups (and grading) of projects done collaboratively, to assigning one grade to all members of a team working on a project, to sharing individually taken test grades – that is, having each member of a working group assigned the average test score of the members of the working group. This latter practice is designed to induce the strong members of a working group help the weaker ones, with the assumed consequences being that not only will the students being tutored by their peers benefit from the help they receive, but that the ones doing the tutoring will benefit from doing the tutoring. It is also used as a vehicle for engendering "esprit de corps" in working groups. Faculty tend to hold strong opinions about the use of such procedures. However, there is little solid research about their effects, and such research would be welcome.

At WPI, students in some sections can decide how to apportion credit among themselves for each project, with the default being equal credit. This opens opportunities for students' examination of their own contributions as well as consideration of what constitutes important or unimportant contributions to a complex project. The elaborate support structure for project work (see example 2 in Section 7) reflects the cumulative experience of a long tradition of project-oriented instruction at WPI.

We note that although various projects have tried different reward schemes for individual and collaborative work, research is needed to understand better the effects of such grading schemes.

## Issues of Reliability, Validity, and Fairness

As put by Harnisch and Mabry (1992), "a superb education might be poorly assessed by what psycho-metricians consider to be a valid test." As we observed in Section 4 of this report, issues of reliability and validity do not disappear when alternative forms of assessment are employed. Rather, their resolution takes different, broader forms. One must worry more, when an assessment is based on fewer but much richer activities, about issues of inter-rater reliability. (Would two independent scorers give the same score? Can scoring be made consistent, if not objective?) And, one must be concerned about whether the activities fairly and accurately represent the more complex objectives of the course – does the activity really test understanding of a concept or a strategy or only a very particular version of it? To a significant extent, solid answers to such questions are simply not yet available, especially if the assessment tasks engage students with complex and multirepresentational concepts such as that of derivative, or in complex activities such as modeling. Hence, for now at least, we must rely on our best collective judgment (which, incidentally, is on what even the classical psychometric definitions of validity ultimately rely). Work by Shavelson and associates at RAND (1989) in the context of elementary school mathematics and science indicates that reliability and validity of alternative assessments are achievable, but only with care and hard work on the part of those who build and evaluate them. In general, these alternative assessments are quite sensitive to variation across content and across students. In particular, background experience of the student and variations in the form of the task have large impacts on outcomes. Also, correlations with traditional measures of ability are less stable across tasks than for traditional short-answer tests.

Again, fairness is an issue. One must consider whether students have had an opportunity to learn such skills as technical writing within the course in which that skill is an important component, or if they have adequate and equal access to the technology that the course assumes to be available. Other aspects of fairness arise in the contexts of group work and how it is graded. Each project has struggled with these matters, and it may be some years before a range of adequate solutions will be constructed. But it is clear that these issues should be dealt with in a careful and sensitive manner, with resolutions varying across differing campus circumstances. It is useful to recall, however, that such matters of reliability, validity, and fairness were relevant even under

traditional instructional and assessment situations, whether or not they had been dealt with explicitly and in depth – they are not new, but they are more important now than ever before.

Costs and Other Constraints

Just as more ambitious curriculum and modes of instruction cost more to deliver than large lecture courses centered on basic computational skills, so do more elaborate forms of assessment. Perhaps the most expensive means of assessment is the structured individual interview, which has been used (mainly as a research tool) in the Purdue Project. However, even laboratory reports and project reports must be graded if they are to be taken seriously by students. And, if they are at the core of the activity, as they are in several projects, then students must be given feedback during the preparation of the project report. At Worcester Polytechnic Institute, for example, for a typical three-week project, students (normally in small groups) bring in a draft report after two weeks for a 15-30 minute meeting with the instructor. On occasion, students meet with more experienced technical writing majors to plan or discuss their project reports. At Duke and elsewhere, the school writing center becomes involved as a resource.

Beyond the additional labor costs of assessments that involve significant amounts of writing (e.g., labs and projects, essay exams, and take-home exams), there are other constraints to be considered. For example, Project CALC discovered that student writing suffered greatly under the time pressure of traditional exams. Their exams are sometimes offered in the evenings, where several hours are available if needed. The limited availability of computers can likewise constrain those who wish to give exams involving computers. Some projects allow study groups as large as seven students to hand in a single assignment. This cuts down on assessment and equipment costs, but with an obvious decrease in individual feedback and involvement. Some constraints have been addressed through the use of take-home exams, which at some projects, (e.g., Calculus in Context) are even used as part of the final exam. Incidentally, just as assessment signals what instructors take seriously, so too does a college or university's willingness to treat as acceptable the costs of more expensive calculus instruction and assessment.

# 6. A Research and Development Agenda for Calculus Assessment

The introduction to this report noted that there have been two major sources of change regarding assessment in calculus: the reform movement itself, which highlights new goals for student attainment, and the fact that a significant part of the reform movement is grounded in a growing understanding of student thinking and learning. The role of research into student thinking and learning, and details pertaining to that research, have been largely implicit in the discussions to this point. Here, before laying out the dimensions of an agenda for calculus assessment, we wish to make the discussion of research explicit. We shall indicate the kinds of understandings that the research provides, and the ways in which extant research can undergird assessment (and therefore instruction). Needless to say, the potential scope of such research is tremendous; a broad review, cursory or not, would cover a great deal of territory superficially and at breakneck speed. We have chosen instead to give a brief, selective discussion of one small but crucially important arena: students' understandings of some elementary aspects of the concept of function. One can extrapolate from the details given here to imagine comparably careful studies of student understanding in all aspects of calculus.

## 6.1 An Illustrative Discussion: Student Understanding of the Concept of Function

There is a large, though hardly comprehensive, amount of research on student understanding of the function concept; see, for example, Thompson (1994) for a selective review of themes related to the topic with a focus at the college level, or Leinhardt, Zaslavsky, and Stein (1990) for a more comprehensive review of major themes in the literature. Here we present three themes and illustrative examples, including suggestions of assessment items to get at some of the understandings they highlight.

    A. *The idea of concept image versus concept definition (drawn largely from the work of Tall, Vinner, and Dreyfus).*

What the mathematician defines as "function" and what students understand by the term can be radically different. Typically, students construct their understandings by making abstractions based on their experiences with functions, and the abstractions they build may or may not correspond to the formal mathematical notions underlying what they have studied. Students may, for example, believe that all functions are continuous; indeed, many students may think that functions must obey very strict constraints (although they may not necessarily be able to articulate their belief) .

For example, it has been shown (Breidenbach et al., 1991; Schwingendorf et al., 1992) that students entering college, and even those who have studied college mathematics for several semesters, often insist that an explicit algebraic or trigonometric expression be available before they are willing to say that a situation can be described by a function. This narrow view can lead to difficulty in understanding a number of mathematical topics such as related rates and implicit differentiation in calculus, underlying issues with regard to differential equations, and duality notions in linear algebra. An interesting assessment/research question would concern the relation between difficulties with these topics and such restricted notions of functions.

In other work on student conceptions of functions, Markovitz, Eylon, and Bruckheimer (1983) presented students with sketches of functions such as the one below:

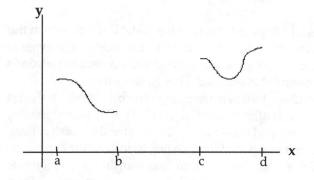

They told the students that the sketch showed part of the graph of the function f, and that f is continuous on the interval [a,d]. They asked students to sketch in values for f on the interval [b,c] so that the result would be continuous. They then asked whether there might be other ways to do so, and if so, how many. A significant proportion of the students completed the graph as follows:

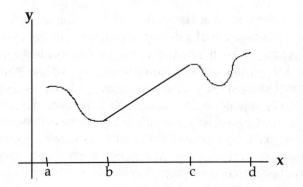

So far, so good. But a significant proportion of those students went on to claim – indeed, they insisted when they were questioned on the issue – that the linear segment joining the point (b,f(b)) with the point (c,f(c)) is the *only* way to extend the function to be continuous on [a,d]!

This is, of course, a fundamental misconception, and one that would cause students subtle but serious conceptual difficulties in their future work. Note that the standard treatments of continuity on traditional calculus exams – asking students to provide the definition of a continuous function and examples of continuous or discontinuous functions, and even to prove that a particular function is continuous at a particular value of x – all leave this misconception unaddressed.

   B.  *Function as action, as process, as object (drawn largely from the work of Dubinsky, Schoenfeld, Sfard, and their colleagues).*

As Thompson notes (1994, p. 26), "It is well known that elementary school students have difficulty conceiving of elementary arithmetical expressions as anything beyond a command to calculate. They typically do not think of, say, 4(12 - (4 + 5)) as representing a number. Similarly, algebra students often think of, say, x(12 - (x + 5)) as representing a command to calculate." By the time they reach college, even this aspect of function may be lost and the expression $7x - x^2$ may be seen as no more than the result of forming a string of symbols obeying certain syntactic rules. Thus,

for such students the meaning of differentiating such an expression is totally contained in the action of dropping x in 7x and moving the 2 in $x^2$ from the exponent to the coefficient while reducing it by one in the exponent.

As Thompson points out, the idea of an expression as a command to calculate "captures some of the 'action' components of functions but leaves some other, critically important aspects unaddressed. Two of the most critical cases are the ability to understand 'function as process (or transformation)' and function as object."

In the former case, the function f given by f(x) = 2x + sin $x^2$ is interpreted as the process that, when given the input x, produces the output f(x). Through abstraction, this goes beyond explicit expressions and can be extended to more complex processes such as those connected with forming the function g given by

$$g(x) = \int_0^x (2t + \sin t^2)\,dt.$$

Here, a successful calculus student might understand the expression as representing a process whereby, given x, one considers a Riemann sum on the interval from 0 to x, refines the subdivision, and passes to a limit to obtain a number. This number is the result of the process. Such a process conception, together with an understanding of composition of functions, might help students understand what happens when g is composed with another function. Eventually, from this point of view, it is possible to obtain an understanding of the Leibniz rule for differentiating a function given by an integral with variable endpoints.

In contrast, there is the notion of function as object – a "thing" independent of what it does, and upon which we act holistically. Thus we can say something about the derivative of a function given by f(ax + b) without knowing much about the derivative of f.

Operations on functions such as translation, composition, differentiation, and integration depend on this notion. So too does the idea of a function being a solution to a differential equation. Likewise, being able to understand families of functions and the behavior of parametrized families of functions (e.g., understanding how the graph of f(mx + b) varies as one varies m or b) depends on the ability to think of any particular member of that family as an object (see, e.g., Ayers et al., 1988).

Being able to move freely between the process and object conceptions of function – knowing when either conception might be profitable for understanding a situation at hand, and exploiting that conception appropriately—is something that people who are mathematically accomplished do naturally, in the sense that it seems effortless. However, that naturalness is learned, and it can be extremely difficult for students. Some mathematics educators (e.g., Sfard, 1991) have expressed the opinion that it may only be possible for a small percentage of students. Others (Ayers et al., 1988; Breidenbach et al., 1991) have shown that certain ways of using computers can help. Assessment tools combined with research would be of great use here.

*C. Functions in multiple representations; connections across them; the situations represented by functions (drawn largely from the ideas of Kaput, Schoenfeld, and others).*

Simply put, the effective use of mathematics depends on the ability to represent situations (both pure and applied) in a variety of ways—with verbal, symbolic, graphical, tabular, and algorithmic representations among them. The effective use of mathematics also depends on being able to interpret the information contained in all of those representations and to move flexibly among them when one representation is more suited than another to provide or communicate information. (The analysis can and should be qualitative as well as purely quantitative.) Here we present two illustrative problems that can be used for purposes of such assessment. In the interest of brevity, we will not provide extended analyses of what it takes to understand and solve the problems.

The first problem (modified from *The Language of Functions and Graphs* (Swan, 1985)) asks for an interpretation of the following graph:

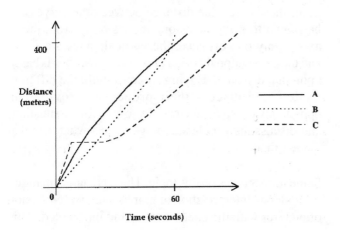

THE HURDLES RACE

The rough sketch graph shown above describes what happens when 3 athletes A, B, and C enter a 400 meters hurdle race.

Imagine that you are the race commentator. Describe what is happening as carefully as you can. You do not need to measure anything accurately.

This problem is mathematically straightforward, and to many faculty it will seem trivial. Yet, student performance on the problem is often quite revealing. In standard curricula, the primary if not sole action students are asked to perform with regard to graphs is to produce them — typically by plotting points and/or sketching the graphs of symbolic expressions. While students may have had extensive experience graphing, they have rarely had experience interpreting graphs. As a result, they may not be attuned to, or may need to think through, the key features of the graph in the hurdles race — e.g., the segment of the race where the graph of runner C's trajectory is horizontal, or the interpretation of what happens close to the finish line.

The part of the race where C's trajectory is horizontal serves as a diagnostic point for one common student difficulty. Many students exhibit confusion about the interpretation of the graph of a function that is constant over an interval: They have difficulty understanding that a graph whose dependent variable is not changing still represents *something*. ("I can see the graph, so something must be happening.") Once such issues are addressed and out in the open, they can be easily dealt with. But having students work a problem such as Hurdles Race, and discuss it in the classroom, can provide instructors with a good idea of what their students actually do know — and the instructors can then tailor their comments accordingly. The skills required for this problem, though elementary, are necessary for interpreting the graphs produced by mathematical models, which students are encountering with increasing frequency as mathematics curricula change.

Readers will have noted the similarities between this problem and the report's first example, which dealt with interpreting distance-versus-time graphs. As exercises to test understanding, these tasks have some interest; if they become generic, there is of course some danger that the skill they call for, interpreting graphs to analytic ideas about a function, could degenerate into a pure routine. However, the skill involved, even if made routine, is still valuable — as long as it is recognized that it is only one of a number of translation skills that compose mathematical competency. There are many such translations (graph to story, table to function, and so on), and a student who is competent in all of them has surely advanced a long way toward understanding. Indeed, having students work on the wide variety of translations between representations is important in general, and it is a specific goal of some of the calculus projects.

The second problem deals with the graphs of functions given analytically and introduces issues related to what is "really" on a graph in comparison to what we perceive to be on the graph. Students are given (or — better — asked to use a graphing package or graphing calculator to produce) the graphs of the functions $y = x^2$ and $y = x^2 + 10$ on an interval that includes x-values from -10 to 10. A graph of those two functions is given in the accompanying figure.

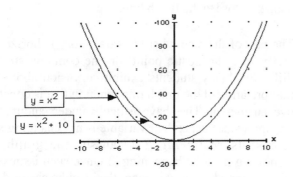

The graphs of $y = x^2$ and $y = x^2 + 10$

Perceptually, the two graphs appear to be closer together for larger |x|. Are they? In what senses of "parallel" are these two curves parallel, and in what senses are they not parallel?

This problem is quite different in its potential student response and its pedagogical impact depending on whether it appears before or after students have dealt with derivatives as slopes. Before derivatives, it helps set the stage for the idea of tangent lines and parallelism of curves. After derivatives, it helps expose different conceptions of parallelism and how they relate to the conventional one based on derivatives. It also reveals weaknesses in most students' conception of parallelism (Schoenfeld, Smith, and Arcavi, 1994). Most students find it easy to agree that the derivative gives the slope of

a curve, that the derivative of a constant is zero, and hence that two functions that differ by a constant must have the same slope. And if they have the same slope, they must be parallel. Moreover, the vertical distance between the two curves is a constant for every value of x. Nonetheless, the mathematics is belied by their (and our) perceptions.

The perception that the curves get closer together as |x| increases can be given mathematical support in a variety of ways, each of which gives the students the opportunity to do some interesting mathematics. Asking the students to do so can provide them with the opportunity to understand that the same thing (in this case the distance between two curves, or between a point and a curve) can be characterized in different ways, with different results. For example,

(i)   Students might determine the horizontal distance between the two curves, say between the point $(a, a^2)$ on the lower curve to the point $(b, a^2)$ on the upper curve. They can explore the behavior of the horizontal distance, |a - b|, as the value of "a" gets large.

(ii)  Students can be asked to find a function that gives the distance from the point $(a, a^2)$ to the point where the normal line to $y = x^2$ that passes through that point intersects the curve $y = x^2 + 10$. They can explore the behavior of this function as the value of "a" gets large.

(iii) Given a point $(x, x^2)$ on the lower curve, the students can be asked to find the coordinates of the closest point on the upper curve and determine the distance to it.

In all these cases, the distance between the curves can be shown to decrease as one moves away from the y-axis. Only one distance, the vertical, stays the same, and the conventional notion of parallel functions is based upon that distance. Having students build the different models and discuss their properties provides an opportunity to explore their ability to model such situations and discuss their understandings of just what it is that models model.

Some further notes on this topic: There is a large literature on students' interpretation of graphs and their confusion of pictures with the phenomena being interpreted. One collection (see Janvier, 1987) offers numerous examples that could easily be tailored for purposes of assessment. There is similar work in physics (e.g., Lillian McDermott's), where students are confronted with velocity or acceleration graphs and asked to describe other aspects of the objects' motion. Generally speaking, many problems that are "cookbook computations" in symbolic form become challenges when presented in other ways. If you present graphs of f(x), g(x), f′(x), and g′(x), and ask questions about h′(x), where h(x) = f(g(x)), students who are fluent at the chain rule as a computational device may well find themselves stymied. For a general discussion of the issue of moving flexibly across representations and from the process to the object perspective, see Moschkovich, Schoenfeld, and Arcavi (1993).

Of course, this collection of examples exploring aspects of student understandings of function barely scratches the surface of what needs to be explored and understood. We have not, for example, discussed operations on functions (such as composition), or summation, or limits. We have not included examples of complex mathematical modeling, for example, or discussions of how students should attend to the relationships between the mathematics of the models they create and the characteristics of the situations they model. These and many more examples all fit under the aegis of assessing students' understanding of function. The examples discussed here are intended to be suggestive – to illustrate the point that while the mathematics may seem straightforward (to those of us who understand it), it is itself quite complex, and that student understanding of that mathematics is far more complex. If we are to make progress in understanding student learning, and in understanding both the effects of our instruction and how to improve it, we will need to confront that complexity. The field would profit from having a research support structure of richness comparable to that described here, for all of the major concepts and issues in calculus. A discussion of useful directions for research and development follows.

## 6.2 The R&D Agenda

The working group has identified seven categories of assessment-related research and development that will provide valuable information regarding the assessment of teaching and learning in calculus. The categories reflect the diverse roles of assessment, as elaborated below.

**Category 1: Combined R&D into (a) understanding student understanding and (b) generating relevant assessment items in concert with that understanding**

For each important goal related to student understanding – call it X – we need to ask: What do we mean by understanding X, and what methods do we have available for assessing the kinds of understandings that students have? As Section 6.1 makes abundantly clear, understanding student understanding is a decidedly non-trivial matter: what we do know points to the complexity of the issue, and there is much more that we need to know. Just as understanding elementary aspects of the concept of function can be expanded as suggested in Section 6.1, each of the topics discussed in Category 1 requires a similar expansion. And, once we have a good handle on what it means to understand X, we need assessment items that will help us and our students determine how far they are along the path to understanding it.

At minimum, two complementary perspectives on what calculus (or any other mathematical domain) is all about are useful to consider. Those perspectives deal with mathematical topics, and cross-cutting mathematical issues. We take them in turn. The discussion of topics and issues given here can be considered an elaboration of the content and process aspects of the Framework for Balance given in Section 3.

**Category 1A: Assessing student understandings of fundamental mathematical themes and topics in calculus**

The following lists of major topics and themes in calculus are drawn largely from the description of goals in Section 2. The two lists provide alternate ways of conceptualizing aspects of calculus content, which we then expand upon mathematically and cognitively.

Topics

- Functions and their various representations;
- Limits and continuity;
- Parametric representations;
- The derivative as an indicator of rate of change, and as the slope of the tangent line;

- The integral understood as a means of aggregating small quantities, and some applications thereof;

- Symbolic integration;

- The fundamental theorem of calculus;

- Numerical methods; and

- Surface and vector calculus.

## Themes

In calculus, there are themes that are both global and local in character. Global themes, that is, themes that pervade calculus (indeed, mathematics in general), have to do with notions such as *mathematizing* – dealing with complex situations by building mathematical representations or models of them, in a close dialectic between the formal mathematics and the situations that the formal mathematics represents. More local themes that either typify calculus or occur frequently in it are

- Motion and its characterization;

- Change, as represented by the derivative, differential or difference equations, etc.;

- Accumulation, as in the integral;

- Approximation and error estimation in differentiation, integration, limits, modeling, . . . ;

- Prediction; and

- Control.

## Mathematical aspects

Each of the topics and themes listed above can be elaborated in terms of the typical mathematical ideas brought to mind when one invokes them:

- Definitions (formal and informal);

- Classes of examples (introductory, illustrative, exemplary, pathological);

- Properties;

- Important theorems and results; and

- Applications.

This list of mathematical aspects just described provides a way of fleshing out a mathematical topic or theme – of

describing the character of the mathematics related to that topic or theme that one would like students to know.

Using the list in heuristic fashion for any particular idea produces a description of an important body of mathematics – e.g., a characterization of what a student is expected to understand about the mathematics of function, or about the ways in which approximation is an important theme that permeates calculus and its applications. Examples of such ideas might be as follows:

- How should students be able to characterize this notion?

- What examples should they be able to produce or analyze?

- With what properties, results, and applications do we expect them to be familiar?

## Cognitive aspects

Now, from the point of view of assessment, we need to examine each of these issues from the cognitive perspective: What does it take to understand the mathematics as characterized above?

- What factual knowledge do students need?

- What representations do they need to have/ use?

- What kinds of visualization are required?

- What mental constructions do students need to make in order to be able to understand it?

- What conceptual understandings must they possess to use the notion (e.g., function as object to understand composition)?

- What connections do they need to make (e.g., understanding function composition to make sense of the chain rule)?

- To what degree is it necessary for them to have certain processes (e.g., symbol manipulation) well learned, so that absence of such performance is not an obstacle?

In addition,

- What are typical student understandings of this topic – typical difficulties, roadblocks, partial understandings, and misconceptions?

With this discussion as background, we are in possession of the elements of a framework for the assessment of students' content knowledge. One can pick a major topic or theme and flesh it out according to its mathematical aspects. One can then select some part of that mathematical characterization and examine its cognitive aspects: What does it mean to understand this part of this piece of mathematics[4] from the mathematician's point of view, and what kinds of student understandings is one likely to encounter? Then, one can create assessment devices (problems, interviews, or other prompts) that provide information about students' understanding.

## Category 1B: Cross-cutting mathematical issues

Many issues relating to collegiate mathematics do not fall under the classification scheme described in Category 1A, but they are centrally related to major goals for many of the calculus projects. Course design is often predicated on assumptions about basic issues: e.g., the role of symbol manipulation in calculus and the amount of practice required for students to become fluent to the degree deemed necessary; or the degree to which basic skills can be developed in a problem-solving context, or might be more efficiently and effectively developed through practice. Little is known definitively about such issues. The more that can be uncovered, the more solid grounding there will be for improved instruction.

The following are some issues that could be clarified.

### The role of contexts

How important is it that students meet mathematical concepts in the context of real applications? Can pseudo-real applications (for example, real story lines with simplified data) do just as well? Do such application contexts make it easier for the students to recall and understand the mathematical concepts? Do they make the subject more interesting? What kinds of benefits are derivable from experience with applications, and how can they be documented? Much of the calculus reform movement is predicated on the importance of context — indeed, some claim that *mathematizing* is at the core of doing all mathematics, and our curricula should reflect that point of view. It would be helpful to gain a deeper understanding of such issues.

### Access to mathematics

How can we make mathematics more accessible to women and disenfranchised minority groups? What attributes of courses make them more appealing to or enhance the success rates of those groups?

### Communication

How important are verbal and written expression in enhancing mathematical learning? Given the recent increased emphasis on communication in mathematics, it is important to understand the ways in which mathematical communication shapes the development of students' ideas.

### Writing

Writing has — for good reason — received increased emphasis in mathematics instruction in recent years, and it is featured heavily (via lab reports, extended assignments, projects, etc.) in the calculus projects. Yet the precise contributions of writing to mathematical understanding, and ways to enhance those contributions, are not well understood.

### Reading

How important is learning to read mathematics as an ability in its own right? How important is it to learning mathematics generally? In what ways does a reading regimen affect students' mathematical performance with regard to precision or clarity of expression?

### Group work

One rationale for group work is that after graduation, students are likely to be team members; individual work is artificial and does not prepare them for the future. Another is that group work allows students to engage in more difficult and complex tasks than they could individually. A third is that group work, properly employed, fosters individual learning. We need to know more about how to structure groups productively and how to identify effective group exercizes, and to what ends. In addition, we need to know more about reward

---

[4] The phrasing "this part of this piece of mathematics" is deliberate. Note that each of the assessment items concerned with the function concept discussed above only deals with one aspect of what it means to understand "function."

structures in assessment of group work: What reward and activity structures in various situations facilitate the right kinds of interactions and are perceived as equitable?

## Community

How important is a learning community in promoting the learning of mathematics? How big should such communities be? Are particular roles in such communities useful in developing particular kinds of skills or understandings? Does participation in such communities have notable effects on students identified as having high or low ability?

## Background

What skills, background knowledge, and level of conceptual development are necessary to profit from and succeed in a calculus course? Should students have prior experience in algebra? trigonometry? visualization? the function concept? deductive reasoning? data analysis? computer technology (word processing, graphing, programming, hand calculators)? other sciences (Newton's Laws, probability, statistics)? What level of general intellectual development is necessary?

## Intellectual context

How much do students benefit from learning about the intellectual history of mathematics? its relation to the development of western scientific and philosophical thought? its importance in the conceptual framework of other contemporary sciences?

## Mathematics as a profession

How important is it to convey a sense of what mathematicians do, i.e., what sorts of problems they think about?

## Symbol manipulation

How is proficiency in symbol manipulation (algorithmic performance) related to conceptual understanding? This is a question with many parts. For example, what level of proficiency at low-level procedures is required as a base for more complex tasks? What level of symbolic proficiency is necessary to give students enough confidence to be willing to dive into new problems? What is the role of routine manipulation practice in establishing the web of connections among symbols and between the symbols and the related mathematical concepts and applications? For some students there is a significant satisfaction in performing symbolic calculations correctly. How important is this in students' enthusiasm for learning? Is this emotional satisfaction more important for potential mathematicians than others?

**Category 2: Studies of the effects of the uses of various kinds of assessments on the students and faculty involved**

There are strong beliefs in the education community, supported by research (see, e.g., Madaus et al., 1992), that "what you test is what you get." Such beliefs, in the context of calculus reform, have led to the use of a variety of innovative assessment techniques. However, very little is known about the relative merits of the different forms of assessments now being used. Suppose one adopts novel assessment measures in a calculus class (as discussed in Section 4, and exemplified in Section 7). Consider, for example, assessments that emphasize

- student research projects;
- a large amount of writing;
- laboratory or project-based mathematics;
- group work; and
- the use of technology.

What are effective ways of implementing such assessments? What influence do such measures have on the instructors? On the students? On the character of the course as a whole? Are there contexts in which particular approaches seem particularly effective? Are there ways of using particular approaches that seem to produce consistently good results?

**Category 3: The creation of assessments as levers for change**

A number of the issues in Categories 1 and 2 are local or technical in the sense that they focus on particular aspects of student performance, learning, or instruction. There are, however, global issues related to the influence of assessment on instruction. Imagine the existence of widely available, high-quality assessments of calculus learning with a particular kind of face validity – the property that mathematicians who examine them tend to feel that "my students ought to be able to do the things on this test, and if they can't, then there's a problem with what we're teaching, or how we're teaching it." Such assessments should, of course, meet some criteria for balance such as the framework described in Section 3.1 and used in Section 7; they should, indeed, contribute to its refinement. If such assessments existed, they would help establish high standards nationwide, and their presence would serve as a lever for change. (Even if they were not adopted as a whole, the ideas embodied in them would be stimuli for reflection.) Thus it might be worth sponsoring their development.

One step along the same lines would be the creation of a Source Book of Assessment Exemplars for calculus – the calculus analog of MSEB's (1993) *Measuring Up*. Such a collection might be ad hoc, or it might be structured along the lines of the framework discussed in Section 3.1. (One interesting task is to take the same basic mathematical idea and create a family of assessment items related to it, corresponding to different dimensions of the framework: a short item, an essay, a project, modifications of those items that are expected to be done with or without the use of technology, open or closed book, etc.)

### Category 4: Studying the effects of various assessment formats on student performance

As we move from standardized tests to other forms of assessment, we face a host of issues regarding the effects of the form of testing on student performance. For example,

- What are the effects of time pressure? Of removing it by giving tests without tightly constrained time limits?

- What are the effects of group work? What does group work reveal about student potential that was hitherto hidden, and what does it obscure?

- What are the effects of open-book versus closed-book tests?

- What are the effects of having students use standard computer packages? Of writing programs themselves?

We note that in both Categories 3 and 4, it would be useful to include studies with planned variation: using the same assessments at multiple sites, or multiple assessments (in different class sections) at the same site, to help tease out the effects of the particular assessments.

### Category 5: The adaptation of known assessment types to calculus assessment

While Category 1 focused on form, this category focuses on content. Simply put, there is no need to reinvent assessment; as discussed above, there are a variety of assessment techniques that have been developed for K-12 mathematics. The right kinds of teams, presumably consisting of people who have knowledge of the assessment techniques working with people who know the mathematics, might find ready adaptations of known techniques.

Candidates might be

- the "construct a test" test;

- having students create concept maps;

- portfolio development; or

- other techniques discussed in Section 4.

### Category 6: Creating assessments that are consistent with expected student uses of mathematics, and the conditions under which they might be expected learn it and use it

Rather than *the* standard calculus course, we face over the next decade a proliferation of calculus courses – some that have a heavy emphasis on modeling, some that make extensive use of technology, and so on. Some of the motivations for such courses will be the students' ultimate use of the mathematics; e.g., potential engineers are likely to work in tool-rich environments, where hackwork computations are done by computers. Assessments should be developed that are consistent with such emphases; e.g., there should be a collection of tool-rich assessments predicated on the assumption that students will be expected to use technologically based mathematical tools.

At the same time, it is important to trace the effects of new forms of calculus instruction on students' subsequent mathematical performance in future courses and in their careers. It would help to address questions such as: What conceptual base will provide students with the appropriate foundations for the effective uses of mathematics in other majors? What is the effect of new forms of calculus instruction on student understanding and performance in other fields?

Studies along the lines suggested in this section would provide (a) better tools for assessing student learning in calculus, (b) deeper understandings of the assessment process itself, and (c) guideposts for the improvement of calculus instruction.

# 7. Assessment Examples

This final section of the report presents a series of assessment items representing some of the assessment work done by calculus projects. In the first two examples we structure our commentary along the dimensions of the Framework for Balance (cf. Section 3). Having illustrated its use, we proceed less formally with the remaining examples – but suggest that readers consider each assessment exercise in the light of the framework.

For purposes of contrast, we first consider a very narrow assessment task before analyzing richer assessment activities. The progression of activities moves toward items of greater scope. Four examples of extended activities (one project, two laboratories, and one extended instructional/laboratory sequence) are discussed. They are followed by two cumulative assessments, a sample end-of-term examination, and a sample two-day take-home examination.

## Example 1: A Computation

Determine $f'(x)$ where $f(x) = x^2 e^{(\cos(x^3 + 5x + 3))}$

## Discussion

This is a routine computation of the sort that has tended to dominate traditional assessments – and hence student activity. It is computationally more complex than is likely to occur in most applications, but its existence is usually justified on pedagogical grounds. It forces systematic attention to the differentiation rules and demands good algebraic parsing skill. It is worth noting, however, that this same parsing skill is necessary for the student to input this function to a typical symbol manipulator in order to operate on it, and then, again, to interpret the output.

### 1A. Implied philosophical and pedagogical goals.

It is, of course, inappropriate to infer broad philosophical and pedagogical entailments from any single example.

If the vast majority of tasks students face are of this type, they will draw one kind of conclusion about the character of calculus and about what is expected of them. If, on the other hand, they encounter such problems as a "skills component" of a course that includes many other kinds of assessment tasks (as in the use of "gateways tests" by Project CALC; see below), then the conclusions they draw will be quite different. A rough analogy follows: practicing scales on the piano helps pianists to develop important skills. But how much of a music assessment should consist of testing on scales, and how much should consist of more musical aspects of piano playing? And what conclusions about the enterprise would music students draw if playing scales constituted the bulk of an assessment?

### 1B. Content.

Obviously, this task involves rules for differentiating specific functions, as well as the product and chain rules, strictly in algebraic form. The expected student difficulty is with adequate application of the chain rule. It is important to note what this task does NOT involve. It does not either develop or reinforce any of the major concepts of the calculus, including and especially the complex notion of derivative. It is unirepresentational, strictly algebraic. As given, there are no connections with either applications or other mathematics. The notion of function-composition underlying the chain rule need be understood only as a routine for substituting symbol-complexes into other symbol-complexes. The entire operation of differentiation tends to be experienced by students as an algebraic operation, which, of course, it is – an operation on character strings. But unlike those for whom this algebraic operation takes place in a rich web of concepts, for students without these concepts the operation is exactly equivalent to fraction simplification, expanding powers of a polynomial, or even solving an algebraic equation – that is, a manipulation of character strings.

1C. Thinking processes.

Of the following list of thinking processes, the problem calls for essentially none: explore, experiment, investigate, formulate, conjecture, hypothesize, analyze, interpret, evaluate, relate, compare, choose, decide, plan, organize, design, make, argue, justify, prove, generalize, recognize patterns, predict, reflect, explain, summarize. Now it is certainly possible to create differentiation activities that involve these processes. See, for example, the part of Example 4 that involves a laboratory exploration of some properties of trigonometric functions.

1D. Products.

The only product is the sequence of computations, which implicitly carries information regarding the reasoning used by the student.

1E. The kind of situation involved (pure/ application/real).

This problem is neither pure nor applied mathematics.

1F. Issues of diversity and access.

Due to the context-free nature of the task, issues of language, context, and cultural content are minimal. This fact points to the opposite tendency associated with a move to highly specific contexts for the development of concepts or strategies, where the risks of disenfranchising students because of cultural bias or the students' limited linguistic or cultural access to such contexts can be much greater.

1G. Circumstances of performance.

Here we are not told of circumstances under which the task is done, e.g., time limitations, but we can safely assume that a symbol-processing technological tool is not available, and the task is to be done by an individual working without help. It may or may not be open book.

1H. Aesthetic issues.

This is not an especially engaging task and is not intended to be — for the performer, this task is akin to the "compulsories" in gymnastics or diving. Creativity and interpretive inventiveness are not the issue.

The Alternative Uses of Standard Computational Examples

Conventional assessment items such as Example 1 have their place, of course, since students need to acquire some level of symbolic proficiency. Many students, especially those who have been taught that mathematics is primarily a matter of symbolic manipulation, feel disoriented if no such problems are assigned or if no such tests are given. To fill this void without giving the impression that symbolic proficiency is the object of the course, various strategies have been used. As mentioned on page 22, such tests are given at Worcester Polytechnic Institute. They can be repeated with only minor penalties until they are passed, but they can affect at most 20 percent of a student's final grade. At Project CALC, they take the form of gateway tests, which may be repeated, but which must ultimately be passed at a very high level (95 percent). Although one cannot receive credit for the course until this requirement is met, the gateway tests do not otherwise affect a student's final grade.

Approaches such as this not only put mechanical computation in its proper perspective, but they also provide tools for diagnosing and correcting computational difficulties. As students take successive tests, their difficulties are identified and discussed with them. The extent of these difficulties shrinks, and the students finish the process with confidence in their computational abilities.

## Example 2.  A Project: Space Probe Rescue

This project assignment comes from a calculus course at Worcester Polytechnic Institute, which has a long tradition of extended and collaborative student projects. As Example 2 exhibit, we provide a slightly edited version of the project specifications, which are extensive, so that readers can see the kind of work that goes into setting up the projects — including the scheduling of progress reports, submission dates for draft reports, and help sessions for writing. Also important is that the scoring scheme is communicated from the beginning: students know the criteria by which they will be graded. A sample student report follows the assignment. It is worth noting the quality of the writing, in contrast to that seen on typical examinations.

Since this assignment comes early in the term, it is relatively simple and is of limited mathematical scope.

Experience suggests that in setting up projects, it is useful to increase complexity and widen the exploration space gradually. Such a measured approach allows students to develop and to build on the skills of exploration, to gain familiarity with tools and support structures, and to learn the skills required to do cooperative work. Thus, in the project at hand, some potential avenues were not examined – chief among these, perhaps, the matter of accuracy of the computations and the relevance of the number of decimal places in the numbers computed. However, faculty report that some groups of students mentioned the fact that straight-line motion was unrealistic because the trajectories would be perturbed by the earth, sun, and other celestial bodies. Another interesting comment came from a group that calculated the time for the probe to reach the Oort cloud as some 8,000 years. They recommended not fixing the probe, but instead waiting for faster ships to become available! Note that the students, whose report is included, not only noticed the relative insignificance of the difference between the two spacecraft arrival times, but that they used this observation to make judgments about what choices to make. In some sense, these judgments were among the more important acts of the modeling exercise.

EXAMPLE 2 exhibit

## Guidelines for Projects and Reports[1]

**Schedule.**

1. January 21. Problem statement for project handed out.

2. January 28. A one-page progress report listing your team members and summarizing your work so far is due in class. This is for your own benefit. It will not be graded as such, but it will reflect negatively on your final grade if you don't hand it in. Also, list times on Thursday and Friday, Feb. 9-10, that you can meet with the peer tutor for a half-hour help session.

3. February 5. The draft project reports are due. You should strive to make the draft as complete and polished as possible, to minimize the amount of rewriting.

4. February 8. The draft reports will be returned to you with comments and suggestions for improvement.

5. February 9-10. Scheduled half-hour help sessions with WPI Technical Writing major or your instructor.

6. February 11. Revised reports due. Your grade on this project will be based on this report and not on the draft report.

**Groups.**

You are to work in groups of at least three and at most four students. Group members are responsible for contributing to the work on the project. Remember that you are responsible for apportioning the grade given on the report among the members of your project team. The way this will be done this term is as follows. If you want grades to be evenly apportioned, you need do nothing. On the other hand, if you want to distribute the grades unequally, the sheet giving the distribution must be signed by every group member.

**Technical Reports.**

A technical report is intended to provide information to the reader. It is not an essay. You should strive for directness, simplicity, and readability. Write your report as if you were explaining your work to other students. That is, you must not assume that your readers already know all about the subject. It is up to you to provide readers with the information they need to understand what you have done and why.

---

[1] This is a slightly modified version of the original document from Worcester Polytechnic Institute.

EXAMPLE 2 exhibit -- continued

**Structure of the Report.**

Your report should be organized around the following sections. Aim your report at explaining the project and your work on it to other students. I am not the audience for your reports. Your project report should contain enough information so that a group of students with similar backgrounds could duplicate your work.

I  Introduction.

This part of the report should state clearly what the project is about, including a brief statement of the problem and a summary of your work. This part should contain enough information to give a reader an idea of what you did. It should also describe the structure of the rest of the report, so that a reader interested in specific information knows where to find it.

II  Background.

This section should contain a more detailed description of the problem. It should also describe alternative approaches you considered in tackling the problem and give reasons why you chose the approach you did.

III  Procedure.

This section should tell the reader what you did. It should contain any data you obtained and describe any references you consulted. It should be detailed enough that another group could duplicate your work. This is also the place where you give the details of how you derived your formula relating time elapsed to the counter reading.

IV  Results and Conclusions.

In this section, you should compare the experimental results with the predictions of your formula. This is also the section where your mathematical model should be described. You should also criticize your model, and make suggestions for its improvement.

EXAMPLE 2 exhibit -- continued

**This form was used to provide feedback on drafts of project reports.**

Group number: _____

Group members:


    Below, you will see some comments on how I think your project report can be improved.  You will

also see numerical scores broken down for each section.  These numbers are intended to give you an

idea of what grade your report would receive if it were handed in unchanged.  As discussed in the

handout, however, only the final grade on your report counts.  There are also some comments from the

peer writing advisor attached.

| Area | | Pts. | Pts. poss. |
|---|---|---|---|
| Organization, | Introduction | | 15 |
| Readability, | Background | | 15 |
| and | Procedure | | 25 |
| Content | Conclusions | | 25 |
| Creativity | | | 10 |
| Assumptions | | | 10 |
| Total | | | 100 |

**Comments.**


        (space left for comments)



(A similar form was used to report final scores.  On that form, instructions for converting the group

score to individual scores were also given. The default was that each student in the group would receive

the score that had been assigned to the group. However, if there was a group consensus, alternate

allocations of the group score were possible.)

EXAMPLE 2 exhibit -- continued

## SPACE PROBE RESCUE

It is the year 2020 and the space probe *Galileo* has been on its way to the Oort cloud for exactly a year. For the last six months, scientists at UNSA (the United Nations Space Authority) have been trying, without success, to free the stuck transmission antenna on *Galileo*. UNSA has announced that a manned rescue mission will be mounted to repair the antenna. You have been called on to do some preliminary feasibility calculations. UNSA has provided you with the following data.

1. The space probe *Galileo* has been traveling in a straight line away from the earth at a velocity of 20,000 km/hr ever since it was launched.

2. Due to budgetary constraints and development problems, the only spacecraft that is capable of such a mission is the *Riemann*. This fusion-powered craft has nearly unlimited range but its engines can only deliver a thrust of 0.1g. Furthermore, the engine cannot be throttled; it is either on or off. There are, of course, small steering engines that can be used to turn the *Riemann* around so that its engine can be used for braking.

3. UNSA estimates that in a year the spacecraft *Euler* will be ready. This craft is essentially similar to the *Riemann*, but its engines will be able to deliver a thrust of 0.2g.

The UNSA would like you to estimate the time it will take for *Riemann* to catch up to and match velocities with *Galileo*. To simplify your calculations, you may treat the problem as motion in one dimension and assume that *Riemann* will be launched exactly one year after *Galileo*. Your results should include a time when *Riemann* should start braking in order to match velocities with *Galileo* just as it catches up. To minimize the duration of the mission, UNSA requests that *Riemann's* engine should be on until you catch up to the probe.

UNSA would also like you to repeat your calculation for the *Euler*, assuming it can be launched exactly two years after *Galileo* was launched. This would give the scientists more time to try to fix *Galileo* from earth.

Your report to UNSA should at least address the following points.

1. Why is a braking period required? That is, what mathematical theorem says that *Riemann* will have to be going faster than *Galileo* for some portion of the mission?

2. Is it possible to obtain an analytical solution for both the time it will take for *Riemann* to match velocities with *Galileo* and the time *Riemann* should be braking? UNSA is very interested in this kind of result, since it will be useful if plans have to be changed.

3. Will the increased thrust of *Euler* make a shorter mission time possible? If not, what thrust would *Euler* have to be capable of to make a shorter mission possible?

EXAMPLE 2 exhibit -- continued

# U.N.S.A

# MISSION

# REPORT

EXAMPLE 2 exhibit -- continued

# SPACE  PROBE  RESCUE

## INTRODUCTION:

There is a space probe directed towards the Oort cloud called the *Galileo*, with which we have encountered some difficulty in the transmission antenna of our *Galileo* space probe directed towards the Oort cloud. We have spent the previous six months unsuccessfully attempting to free the antenna. This report is announcing a manned rescue to repair the faulty antenna. From the proposed crafts, the *Riemann* was our choice due to its low cost and possible high efficiency in fixing the *Galileo*.

The experts' task was to calculate the needed acceleration, distance, and velocity of each ship and have the *Riemann* meet the *Galileo* in the shortest, most effective way. Knowing that the *Riemann* is a nuclear spacecraft capable of accelerating at a steady .1g rate for an indefinite period of time, they found the time for the *Riemann* to meet the *Galileo* at 236.5649 hr. Turn around time was 119. Ø698 hr. They were also given the information that in a year's time, another craft, the *Euler*, will be ready. This craft is identical to the *Riemann* except that it delivers a .2g thrust. They found that the *Euler* would reach the *Galileo* in 235.7696 hr. The turn around time for this craft is 118.2785 hr. The following is their report...

EXAMPLE 2 exhibit -- continued

**BACKGROUND:**

We have been following the *Galileo's* mission since its launch back in 2Ø19. From our extensive research we have determined that the *Galileo* has been traveling in a straight path away from earth at a velocity of 2Ø,ØØØ km/hr since its launch a year ago. We have also found the *Riemann* to be a fusion powered craft that has an engine that delivers a .1g thrust. The engine can not be throttled, it is either on or off, but it can be turned around to be used for braking.

Upcoming calculations require understanding of the following variables:

| | |
|---|---|
| t | time |
| tØ | turn around time |
| s(t) | distance of *Galileo* from origin |
| x(t) | distance of *Riemann* from origin |
| SØ | position of *Galileo* at t= Ø |
| VØ | velocity of *Galileo* |
| a(t) | acceleration of *Riemann* |
| A | constant maximum thrust of *Riemann* engine (.1g) |

The acceleration, velocity, and distance had to be calculated for each craft. The *Galileo's* acceleration and velocity were Ø and 2Ø,ØØØ km/hr respectively, while the distance was found by using:

$$ds/dt=VØ \qquad\qquad s(Ø)=SØ$$
$$S(t) = VØ\, t + c$$
$$S(Ø) = c = SØ$$
$$S(t) = VØ\, t + SØ$$

The *Riemann's* formula for velocity is the integral of the acceleration and the distance is the integral of the velocity. We worked out the preceding equations before working with actual numbers involved.

EXAMPLE 2 exhibit -- continued

acceleration $\quad$ a(t) $\quad$ = $\quad$
$$a(t) = \begin{cases} A & \emptyset < t < t\emptyset \\ -A, & t > t\emptyset \end{cases}$$

velocity $\quad$ dv/dt = a (t) $\quad$ V= $\emptyset$

$$V(t) = \int_{\emptyset}^{t} a(s)ds = \begin{cases} \int_{\emptyset}^{t} A \ ds, & \emptyset < t < t\emptyset \\ \int_{\emptyset}^{t\emptyset} A \ ds + \int_{t\emptyset}^{t} -A \ ds, & t > t\emptyset \end{cases}$$

which gives:

$$\begin{cases} At, & \emptyset < t < t\emptyset \\ -At + 2At\emptyset, & t > t\emptyset \end{cases}$$

From this, distance is calculated.

distance = dx/dt = V(t), $\quad$ x($\emptyset$) = $\emptyset$

$$x(t) = \int_{\emptyset}^{t} V(s)ds = \begin{cases} \int_{\emptyset}^{t} As \ ds & \emptyset < t < t\emptyset \\ \int_{\emptyset}^{t\emptyset} As \ ds + \int_{t\emptyset}^{t} (-As + 2At\emptyset)ds, & t > t\emptyset \end{cases}$$

which gives:

$$x(t) = \begin{cases} 1/2 \ At\text{^}2 & \emptyset < t < t\emptyset \\ 1/2 \ At\text{^}2 + 2At\emptyset t - At\emptyset\text{^}2, & t > t\emptyset \end{cases}$$

## PROCEDURE:

To determine numerical values, we requested the assistance of a computer program entitled Maple. This system cuts down on the time spent and human error made dealing with plugging into formulas.

EXAMPLE 2 exhibit -- continued

A basic concept figured into our calculations was the fact that the *Riemann* must go faster than the

*Galileo* at some point to match distances at some point. This was mathematically proven by

knowing that V(ave) t = d.

*Galileo's* Vt must equal *Riemann's* Vt for the distances to be equal. We know that the

*Galileo* has been moving a year longer than the *Riemann* thus:

>    *Galileo's* (t) > *Riemann's* (t)

which means:

>    *Galileo's* (V) < *Riemann's* (V).

When known variables were put into the figured equations previously given in this report,

the following resulted:

The *Galileo*

$s(t) = 1.7532 \times 10^{8} + 20,000 \ t$

The *Riemann*

a(t)    $12700.8$ km/hr$^2$,    $0 < t < t0$
        $-12700.8$ km/hr$^2$,         $t > t0$

x(t)    $6350.4 \ t^2$,                                                    $0 < t < t0$
        $-6350.4 \ t^2 + 25401.6 \ (t0) \ t - 12700.8 \ t0^2$,            $t > t0$

v(t)    $12700.8 \ t$,                          $0 < t < t0$
        $-12700.8 \ t + 25401.6 \ t0$,               $t > t0$

$t0 = 119.0698$ hr

$t = 236.5649$ hr

We plotted the distance graphs for the *Riemann* and *Galileo* and found t as the point where

the two graphs intersect at a single point. The graph can be found in Figure 1. After some

thought, we decided an analytical solution for the time it takes for matched velocities and the time

EXAMPLE 2 exhibit -- continued

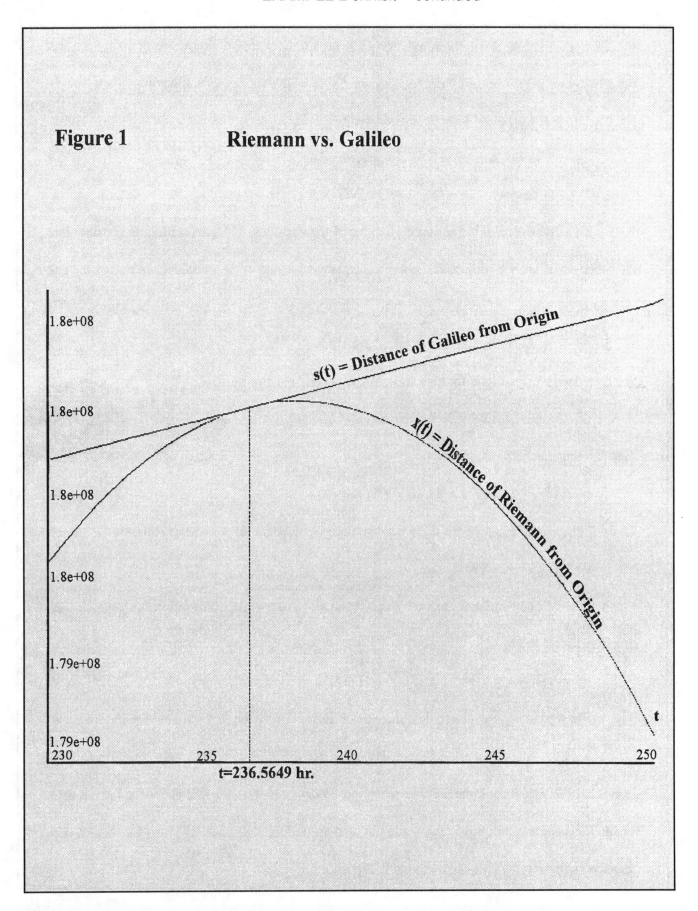

**Figure 1**        **Riemann vs. Galileo**

EXAMPLE 2 exhibit -- continued

the *Riemann* should begin braking was needed in case of unexpected alterations in plans. To find the general equations for t and tØ, we used the knowledge that at time t, the velocity and the distance from the origin of both ships will be equal. Thus,

Distance: $-1/2At^2 + 2AtØt - AtØ^2 = SØ + VØt$

Velocity: $-At + 2 AtØ = VØ$

By solving one of the equations for one of the variables, and substituting this answer into the other equation, we could find the general equation for one of the variables. We used the solve command on Maple to find the equation for t. It turned out to be the quadratic equation:

$A^2t^2-2AVØt-Vo^2-4ASØ=Ø$

We then used the quadratic formula and got the general equation for t:

$t=[2AVØ +[(2AVØ)^2 - 4A^2 (-VØ^2-4ASØ)]^{1/2}] /2A^2$

This formula can be reduced to:

$t=VØ /A + 2[(VØ^2 + 2ASØ)^{1/2}] /2A^2$

To find the general equation for tØ, we substituted our formula for t into the equation

$tØ = (VØ + At) /2A$

which is the velocity equation for both ships set equal to each other. We found the general equation for tØ to be:

$tØ = VØ/A + [(VØ^2 + 2ASØ)^{1/2}] /2A^2$

We were able to evaluate the future possibility of using the *Euler* in place of the *Riemann*. We plugged the appropriate A, SØ, and VØ values for the *Euler* into the above equation and found that the time it took the *Euler* to reach the *Galileo* one year after the proposed launch date for the *Riemann* was 235.7696 hours, and the turn-around time was 118.2785 hours. A plot of distance vs. time showing t can be found in Figure 2.

EXAMPLE 2 exhibit -- continued

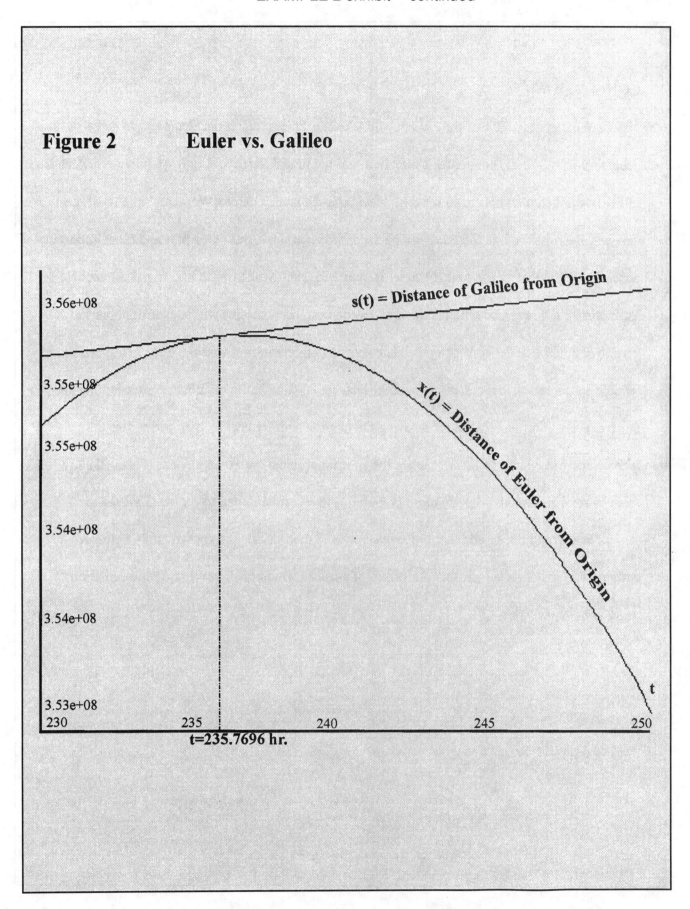

**Figure 2          Euler vs. Galileo**

EXAMPLE 2 exhibit -- continued

**CONCLUSION:**

We can now see that the *Euler* would take less time to reach the *Galileo*. We do feel it's necessary to call to the committee's attention that the time difference is only 48 minutes, which is rather insignificant compared to the time scale between launch dates. We would like to suggest delaying the mission until the *Euler* can be used, for several reasons. Our first and most important consideration is that by the time the *Euler* is launched, Neil Armstrong XXX, our most capable astronaut, will be finished with his therapy sessions and his problems with claustrophobia and vertigo should be under control again. The second consideration is that the *Euler* is a much more reliable spacecraft than the *Riemann*, and the chances of completing the mission successfully are much greater. Our last reason for delaying the mission another year, is that during this time, technicians will be able to further isolate the specific problem with the antenna. This will allow them to send astronauts who are more prepared, and better equipped to fix the *Galileo*.

We would like to add that the *Riemann* would theoretically be able to attain light speed within 4.8 years of its launch, and our group of analysts would gladly evaluate future missions using either spacecraft.

Discussion

### 2A. Implied philosophical and pedagogical goals of Example 2.

From the project description alone, it is clear that projects are important and weigh heavily in student assessment; collaboration is taken seriously; students are expected to work continuously on projects (hence the progress reports); revising one's work is a natural part of the development process (draft reports are examined); writing to communicate clearly is a highly valued activity; mathematical models are important (by virtue of their being the subject of the report); and models (and one's work in general) can be improved on by reflecting upon them.

### 2B, C ,D. Content; Thinking processes; Products.

This project involves a moderate amount of quantitative and spatial reasoning, simple integration, and algebra. The students must relate three related motions and make judgments based on their models of those motions. The problem requests analytical representations and invites, but does not require, coordinate graphical representations. (Numerical representations could have been created as well, but they were not produced by the students whose work is included; similarly, a schematic map could also have been used.) The problem involves building close connections between mathematics and elementary physics, and between mathematics and a situation involving practical and political judgments. Yet, as noted above, some potentially rich mathematical avenues were not pursued.

The construction and discussion of mathematical models, elementary though they may be, still extend significantly beyond the requirements of the standard calculus course; they require skills of formulation, comparison, and justification not called for in routine exercises, as well as the use of technological tools. The write-up clearly focuses on important communication skills absent from the traditional course.

### 2E, F, G, H. The kind of situation involved (pure/application/real); Issues of diversity and access; Circumstances of performance; Aesthetic issues.

This is more an illustrative use of initial value problem situations than a realistic one, at least relative to the motion itself; actual models of the motion of space probes are quite complicated. It is not a "real" problem in the sense that the following problem is:

> Construct a model of traffic flow to describe the way in which a sequence of cars passes through a busy intersection governed by a stop light. Do so by gathering data about the actual flow of cars through the intersection once the light turns green. Determine what the most important variables controlling traffic flow are, and build a model that incorporates those variables. On the basis of the model, and on the basis of practical realities (e.g., how long will people be willing to sit at a red light?), make and justify recommendations for the amount of time the light should remain green (or red) in order to have the most efficient traffic flow (Burkhardt, 1981, pp. 54-60).

Nonetheless, the situation described in the space probe problem is realistic in the sense of involving some model-based decision making, including the inclusion of constraints of various kinds and the need to make practical judgments to ignore differences that are too small to distinguish outcomes.

The task may seem somewhat male-oriented, if only because engineering and space-related tasks currently have such associations in our culture. It appears to be linguistically accessible, and the instructional context – including various support structures for checking progress, getting help with writing and feedback on drafts, having easy access to the technology, etc. – seems designed to keep barriers to a minimum.

## Examples 3 and 4: Two Extended Laboratory Examples

The laboratory projects reproduced in Examples 3 and 4 are from a calculus program developed in Project CALC at Duke University. The program features an integrated weekly two-hour computer lab and emphasizes real-world problems. All laboratory projects are worked on by teams of two or possibly three students. These particular labs required the submission of a formal written report. The assignments and some sample student work are reproduced. Each of the student reports was submitted once, returned with comments by the laboratory instructor, and submitted a second time for a grade.

**Example 3: Falling Bodies with Air Resistance**

This lab occurs in the sixth week of the first semester. The students have been introduced to the notion of an initial value problem and have seen a prototype of Euler's Method for the numerical solution of such problems. This is their first formal exposure to the method.

The students work through the model with no air resistance as a homework problem. In the lab, they are guided through the numerical steps for the model with air resistance proportional to the velocity and then compare this numerical method with the analytical solution. Also, they can check their numerical solution by comparing the apparent terminal velocity with the value obtained directly from the differential equation.

They follow this by setting up (without prompts) the Euler's Method computations for the model with resistance proportional to the square of the velocity. In the second case, they have the same graphical information about the solution as they do in the first case, even though they do not have an explicit formula for this solution.

EXAMPLE 3 exhibit

# Instructions for Lab 6: Falling bodies with air resistance

## Purpose.

We will study "differential-equation-with-initial-value" problems generated by modeling falling bodies subject to air resistance. To investigate these problems we will use Euler's Method, a numerical method for approximating the solutions of such problems.

## Preparation.

Review the discussions in Chapter 2 of the falling body problem (Subsections 2.2.1 and 2.2.2) and of differential-equation-with-initial-value problems (Subsection 2.3.3). Read these instructions carefully, and work the exercise on page 2 before you come to the lab.

When we first considered the speed of a falling body as a function of time, we considered only the force of gravity. This force is directed downwards with a magnitude of $mg$, where $m$ is the mass of the object and $g$ is the acceleration due to gravity, approximately 32.2 feet per second per second. If we let $v$ denote the speed of the falling body in feet per second, then the acceleration is just the derivative of the speed, $\frac{dv}{dt}$. Substituting in Newton's Second Law of Motion, we obtain

$$m \frac{dv}{dt} = mg.$$

When we divide both sides of the equation by $m$, we have $\frac{dv}{dt} = g$. We assume that the initial velocity is 0; that is, the object is dropped. Thus the velocity satisfies the differential equation,

$$\frac{dv}{dt} = g$$

with the initial condition

$$v(0) = 0.$$

It is easy to guess a solution to this problem:

$$v = gt.$$

Since $v$ is also $\frac{ds}{dt}$, this solution to the velocity-from-acceleration problem can be interpreted as another differential-equation-with-initial-value problem:

$$\frac{ds}{dt} = gt$$

EXAMPLE 3 exhibit -- continued

with

$$s(0) = 0.$$

This problem clearly has the solution

$$s = \frac{g}{2} t^2$$

for the distance fallen, $s$, as a function of time $t$. That was our starting point in the text; the constant $c$ that appears there is $\frac{g}{2}$.

In this lab we investigate the corresponding problems that arise when we take into consideration the resisting force of the air through which the object falls. The usual physical assumption is that the resisting force is proportional to some power of the velocity, but the particular power (first, second, other) depends on the particular object. In our case, we'll consider raindrops falling from a cloud 3,000 feet above the ground.

*Exercise. If we ignore air resistance, as we did above, how long (in seconds) would it take a raindrop to fall from a height of 3,000 feet? How fast would it be traveling when it hit the ground? Give your answer first in feet per second, and then convert it to miles per hour.*

If the raindrop is small, say a drop of diameter 0.00025 feet (or 0.003 inches, a size found in a drizzle), the resistive force is proportional to the first power of the velocity; in other words, this force can be described by $F_r = -kv$. Here the minus sign indicates that the force is in the opposite direction to the velocity, i.e., upward rather than the positive downward direction. When we combine this force with the force of gravity, $F_g = mg$, and substitute into Newton's Second Law, we obtain

$$m \frac{dv}{dt} = mg - kv.$$

## EXAMPLE 3 exhibit -- continued

Dividing by $m$ again, we have

$$\frac{dv}{dt} = g - \frac{k}{m}\,v.$$

We'll let $c$ represent the quotient of the two constants $k$ and $m$. When we add in our initial condition, $v(0) = 0$, we have a new differential-equation-with-initial-value problem:

$$\frac{dv}{dt} = g - cv, \quad \text{with} \quad v(0) = 0.$$

Experimental evidence gives an approximate value of 52.6 for $c$.

We want to understand how, under these assumptions, the velocity varies as a function of time. We would like to graph the function, maybe calculate a few values for particular times $t$. Our first impulse is to follow the pattern for the simpler problem: somehow guess a formula for $v$ as a function of $t$, then use the formula to draw the graph and calculate the values. We will begin this lab by taking another approach, one that uses the calculating ability of *MathCAD* in place of the formula.

### Report.

Fill in one copy of these instructions as directed, and turn in that copy with a completed copy of your *MathCAD* worksheet.

### Project.

1. Start *MathCAD*, and load Lab6. Notice that we have assigned $g$ a value of 32.2 and $c$ a value of 52.6. We will calculate *approximate* values for the velocity at $n$ equally spaced points in the interval from 0 to 0.2 seconds. The value of $n$ is assigned as a *global variable* at a convenient spot farther down in the worksheet. The initial value of $n$ is 20. You will notice that the length of the interval between any two consecutive points is denoted by $\delta t$,[1] which is calculated from the formula $\delta t = \frac{0.2}{n}$. The time points themselves, denoted $t_k$ for $k$ varying between 0 and $n$, are determined by the formula $t_k = k\,\delta t$.

The idea is to calculate approximations $v_0, v_1, \ldots, v_n$ to the velocity at $t_0, t_1, \ldots, t_n$. Well, we certainly know what $v_0$ is; the initial condition sets $v_0 = 0$. Fill this in on the worksheet. We also know what the slope of the graph of $v$ versus $t$ is for $t = 0$. This is given by the differential equation; here $\frac{dv}{dt} = g$ [since $v(0) = 0$]. Thus, whatever $v_1$ might be, we know that $\frac{\text{rise}}{\text{run}} \approx \frac{dv}{dt}$, i.e,

$$\frac{v_1 - v_0}{\delta t} \approx g.$$

---

[1] In the text these differences are denoted by $\Delta t$. Since *MathCAD* does not have an upper case delta, we are using the lowercase $\delta$ instead.

## EXAMPLE 3 exhibit -- continued

Since $v_1$ is just supposed to be an approximation to the velocity at $t_1$, we can let it be determined by the formula

$$\frac{v_1 - v_0}{\delta t} = g.$$

Explain here why this implies that $v_1 = v_0 + g\ \delta t$.

Enter this formula for $v_1$ in your *MathCAD* worksheet.

Now we want to calculate the approximation $v_2$ to the velocity at $t_2$. If the graph of $v$ as a function of $t$ passes through $t = t_1$ and $v = v_1$, we know what the slope of the graph is from the differential equation:

$$\frac{dv}{dt} = g - cv_1.$$

Thus, whatever value we select for $v_2$, we want

$$\frac{\text{rise}}{\text{run}} = \frac{v_2 - v_1}{\delta t} \approx g - cv_1.$$

Again, we may define $v_2$ so that

$$\frac{v_2 - v_1}{\delta t} = g - cv_1$$

or

$$v_2 = v_1 + (g - cv_1)\ \delta t.$$

Enter this value for $v_2$ on your *MathCAD* worksheet.

Continue this process to determine a value for $v_3$, and enter that on the worksheet.

It is clear that it is will be tedious to enter definitions for all 20 values of $v_k$ on the worksheet. To make matters worse, we plan to increase $n$ to be several hundred. At this point we need to remember the *recursive* definitions we made in Labs 3 and 4. In Lab 3, for example, we knew the beginning value of the principal and how to make the step from $k$ to $k+1$. We were able to

## EXAMPLE 3 exhibit -- continued

write the instructions for the step down *once* in a recursive definition. Then *MathCAD* figured out how to determine $P_1$ from $P_0$, $P_2$ from $P_1$, and so on.

We have the same situation here. We have a starting value, $v_0 = 0$, and we know how to obtain $v_{k+1}$ from $v_k$. Write down the general recursive formula here, and enter it on the worksheet:

$$v_{k+1} =$$

This method of approximating the solution of an initial value problem by "bootstrapping" forward, using the differential equation and $\text{slope} = \frac{\text{rise}}{\text{run}}$, is called *Euler's Method*. You will probably recognize that we used this idea, without the name, in Lab 4.

2. When you move down the worksheet, you will find a plot box ready to display the calculated values of $v$. Underneath, you will find the global definition[2] of $n$. Display the plot. Move the cursor to the value of $n$, and display the plots for increasing values of $n$ between 20 and 100.

Describe the sequence of plots and how they change as $n$ increases.

Describe how the velocity of the falling object varies as *time* increases.

Especially notice that the velocity seems to approach a limiting value, called the *terminal velocity*. Use your calculated values of $v_k$ to estimate the terminal velocity.

---

[2]*MathCAD* makes an initial pass through the worksheet reading all the global definitions and calculations. Then it makes a second pass reading the conventional ones. Global definitions may be identified by the appearance of the character $\equiv$ in place of $:=$. The $\equiv$ character may be obtained on the *MathCAD* worksheet by typing the $\sim$ character located above the ' key.

EXAMPLE 3 exhibit -- continued

The terminal velocity can be calculated directly from the differential equation. Explain how to make this calculation and then carry it out below. (Hint: What happens to $\frac{dv}{dt}$ as $v$ approaches the terminal velocity?)

3. You will notice that the raindrop approaches the terminal velocity quite rapidly. Estimate the time it takes the drop to fall to the ground from 3,000 feet by assuming that the velocity is the constant terminal velocity the whole time. How does this compare to your answer to the Exercise?

4. For this particular initial value problem, we *can* find a formula for the solution:

$$V(t) = \frac{g}{c}(1 - e^{-ct}).$$

Early in Chapter 3, we will discuss how *you* could find such a solution. For now, enter the definition of $V(t)$ on the worksheet above the plot box for $v_k$, and plot $V(t_k)$ together with $v_k$ in the box. (You will want to change the plot format for one of the functions so that you can tell them apart. Refer to the Instructions for Lab 5 to see how to do this.) Return $n$ to 20, and gradually increase to 100 again. Describe how the exact solution $V(t)$ of the differential-equation-with-initial-value problem and the numerical approximation compare.

## EXAMPLE 3 exhibit -- continued

5. Use the formula for $V(t)$ given in Step 4 to find the terminal velocity. Is this the same as the value you obtained in Step 2? (There is more space on the next page.)

6. For large raindrops, say with diameter 0.004 feet (or 0.05 inches, a size typical of drops in a thunderstorm), the force of air resistance is proportional to the *square* of the velocity. If we use the symbol $w$ for velocity (to keep it distinct from the velocity of the drizzle drop), the differential equation now has the form

$$\frac{dw}{dt} = g - \alpha w^2,$$

where $\alpha$ is another constant. In this case, the experimental evidence yields a value for $\alpha$ of 0.115. Show that this differential equation again implies that the velocity approaches a terminal velocity, i.e., that $w$ has a limiting value as $t$ becomes large, and determine this limiting value.

Again in this case, estimate the time it takes the drop to fall to the ground from 3,000 feet by assuming that the velocity is the constant terminal velocity the whole time.

## EXAMPLE 3 exhibit -- continued

7. We want to approximate the solution to this problem by using Euler's Method. In this case, the velocity takes longer to stabilize, about 2 seconds, rather than 0.2 seconds, as in the previous case. The change in the time scale is already entered into the worksheet. The global variable for the number of calculations in this case is $q$; it is defined under the second plot box.

Move down to the beginning of the second worksheet page, and enter $w_0 = 0$ and the Euler's Method recursive formula for $w_{k+1}$ in terms of $w_k$. Increase $q$ by stages from 20 to 100. Describe the resulting plots of $w_k$ and your corresponding impression of this velocity function. How does it compare to the drizzle situation?

8. Save your worksheet, and print out a copy of the completed worksheet.

COMMENT: In the drizzle-drop problem, with resistance proportional to the first power of the velocity, we had both the Euler's Method approximation and the exact solution. For the thunderstorm-drop problem, with resistance proportional to the square of the velocity, we had only the Euler's Method approximations, yet we obtained the same sort of insight into the solution.

5/19/92

# Lab 6: Falling bodies with air resistance

$g: = 32.2$                                    The acceleration due to gravity

$c: = 52.6$                                    The proportionality constant for the resisting force

$k: = 0..n$                                    The running index for the recursive iteration

$\delta t: = \dfrac{0.2}{n}$                   The length of the time step

$v_0: = 0$                                     Enter the initial condition.

$v_1: = v_0 + g \cdot \delta t$                Enter the first approximate value.

$v_2: = v_1 + [g - c \cdot v_1] \cdot \delta t$   Enter the second approximate value.

$v_{k+1}: = v_k + [g - c \cdot v_k] \cdot \delta t$   Enter the recursive step.

$t_k: = k \cdot \delta t$                      The time steps

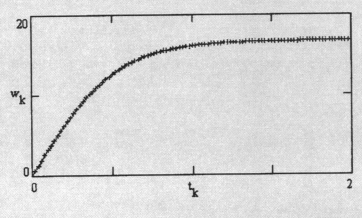

$$w_q = 16.721$$

$n \equiv 100$                      Global definition of the number of steps (first problem)

Student Lab Sheet, Part 1

EXAMPLE 3 exhibit -- continued

**Changes for the second problem:**

$$k : 0..q \qquad \delta t : = \frac{2}{q} \qquad t_k : = k \cdot \delta t \qquad\qquad \text{longer time interval;}$$

$$\alpha : = 0.115 \qquad\qquad \text{new proportionality constant}$$

**Enter Euler's Method formulas here:**

$$w_0 : = 0$$

$$w_{k+1} : = w_k + \left[ g - \alpha \cdot w_k^2 \right] \cdot \delta t$$

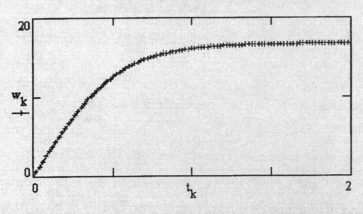

$$w_q = 16.721$$

$$q \equiv 100 \qquad\qquad \text{Global definition of the number of steps (second problem)}$$

Student Lab Sheet, Part 2

EXAMPLE 3 exhibit -- continued

# Report for Laboratory 6

In Lab 6, we studied a differential equation with initial value problem generated by modeling falling raindrops subject to air resistance. In the first part of the lab, we calculated the terminal velocity of small falling raindrops using Euler's Method, which gave us a numerical approximation, and the differential equation, which gave us a direct answer. We then found the exact formula for the velocity and used it to acquire a solution for the terminal velocity. We then compared the various methods for finding the terminal velocity. In the second portion of the lab we approximated the terminal velocity for a model of raindrops of larger size, once again using Euler's Method, and then calculated the terminal velocity directly from the differential equation. We then compared the terminal velocities of the two different sizes of raindrops.

In preparation for the lab, we first calculated the time a raindrop takes to reach the ground from a height of 3,000 feet and its speed when it hits the ground, without accounting for air resistance. We used the equation, $s = \frac{g}{2}t^2$ , for the distance $s$ in feet fallen as a function of time $t$ in seconds, to calculate the time required for the raindrop to fall. We obtained this equation by solving two successive initial value problems.

We first used the expression $\frac{dv}{dt}$ to represent the derivative of the velocity function, $v(t)$, where $v$ is the velocity of the falling object in ft/sec. The derivative of velocity is acceleration, and since acceleration for free fall is that due to gravity, represented by the constant $g$, or about 32.2 ft/sec$^2$, then $\frac{dv}{dt} = g$. This can also be obtained by Newton's Second Law, which states that force equals the product of mass and acceleration, which can be represented as $F = ma$ or $F = m\frac{dv}{dt}$. $m\frac{dv}{dt}$ thus, equals $ma$. If we divide both sides by $m$, we obtain $\frac{dv}{dt} = g$, which is what we found earlier. Given the differential equation $\frac{dv}{dt} = g$ and the initial condition that the raindrop has zero velocity at time 0, or $v(0) = 0$, we found the function that satisfies this initial value problem. Antiderivatives of $\frac{dv}{dt} = g$ have the form $v(t) = gt + C$ by the power rule for antidifferentiation. Since $v(0) = 0$ at time 0, we substituted these values into the antiderivative equation to obtain $0 = g*0 + C = C$. $C$, therefore, had to equal zero, and our solution for the initial value problem had to be $v(t) = gt$.

We subsequently proceeded to find a function for distance. Since $v$ was equal to $\frac{ds}{dt}$, the derivative of the distance function, we rewrote the function $v(t) = gt$ as $\frac{ds}{dt} = gt$. When we combined this differential equation with the initial condition that the raindrop

EXAMPLE 3 exhibit -- continued

has fallen 0 feet at time 0, or $s(t) = 0$, we had the two elements of an initial value problem, and we found its solution. Antiderivatives of $\frac{ds}{dt} = gt$ have the form $s(t) = \frac{g}{2}t^2 + C$ by the power rule for antidifferentiation. Since $s(t) = 0$ at time 0, we substituted these values into the antiderivative equation to obtain $0 = \frac{g}{e}*0 + C = C$. $C$, therefore, had to equal zero, and our solution for the initial value problem had to be $s(t) = \frac{g}{2}t^2$.

Given this function, we calculated the time required for the raindrop to fall 3,000 feet as well as the raindrop's velocity upon striking the ground. We made both of these calculations neglecting the effects of air resistance. If $s(t) = 3,000$, then $\frac{g}{2}t^2 = 3,000$. We divided both sides by $\frac{g}{2}$ to obtain the equation $t^2 = \frac{6,000}{g}$. We then took the principal square root of both sides to find the value of $t$: $t = \sqrt{\frac{6,000}{g}}$, or substituting 32.2 for $g$, about 13.7 seconds. We then plugged this value back into the velocity function to obtain the velocity of the raindrop upon hitting the ground:.
$v(t) = gt = 32.2*13.7 = 441$ ft/sec. We left out units in calculations for simplicity. The values we obtained, also, were approximations.

We then took into account the resistive force of air and proceeded to calculate terminal, or maximum, velocity of the raindrops. We modeled the acceleration of small raindrops (diameter of 0.003 inches) by the function

$$\frac{dv}{dt} = g - cv, \text{ with } v(0) = 0$$

where $v$ is the velocity of the raindrop in ft/sec and $g$ and $c$ are constants. We obtained this function by using the principle that the resistive force of air against a raindrop in a drizzle is proportional to the first power of the velocity of the drop, or $F_r = -kv$. By Newton's Second Law, the force of gravity, $F_g$ equals $mg$. Thus, taking into account air resistance, we constructed the equation $m\frac{dv}{dt} = mg - kv$. Dividing through by $m$, we obtained $\frac{dv}{dt} = g - \frac{k}{m}v$. Substituting in $c$ for $\frac{k}{m}$, we obtained the equation $\frac{dv}{dt} = g - cv$. Experimental evidence has shown $c$ to equal approximately 52.6 reciprocal seconds.

Using Euler's Method, we approximated the values of $v_1$, $v_2$, $v_3$, and finally $v_{k+1}$. We started from the known point $v_0$, and moved to the next one, $v_1$, by computing the rise as slope times run, where the slope is given by the differential equation above, and the run is $\delta t$. At $v_0$, $v(0) = 0$. Thus, our differential equation became $\frac{dv}{dt} = g - (c*0) = g$. The slope at $v_0$, in other words, was $g$. We then multiplied this quantity by $\delta t$, the run, to obtain the rise. We subsequently added this quantity to the value of $v_0$ to obtain an

EXAMPLE 3 exhibit -- continued

approximation for $v_1$. Our equation for calculating $v_1$ thus became $v_1 = v_0 + g\delta t$. We let $\delta t = \dfrac{0.2}{n}$ where $n$ was the number of divisions on the x-axis at which we could measure the value of the function, and set $n$ to be a large number –150– in order to obtain a high degree of accuracy.

We then found a recursive formula for further iterations. We allowed each successive velocity to be represented by the expression $v_{k+1}$. By Euler's Method, this value could be approximated by adding the previous value, represented by $v_k$, to the product of the slope, represented by $g - cv$ (the right-handed side of the differential equation) and the run, represented by $\delta t$. The general formula for calculating successive points thus became

$$v_{k+1} = v_k + (g - cv_k)\sigma t, \ k = 0, 1, 2. \ldots n$$

Once we entered the recursive formula, MathCAD was able to graph the velocities for various times $t$. The x-axis extended from 0 to 0.2, and the y-axis extended from 0 to 0.7. The graph showed that there was a sharp increase in velocity between $t = 0$ and $t = 0.10$ seconds. Then the velocity leveled off to a maximum, or terminal, velocity. We looked at the graph to get a general sense of where this leveling-off occurred. We then chose a high value of $k$ — in this case, 100 — beyond this leveling-off point and had MathCAD provide us the value for at this point. We found that $v_{100} = 0.612$. We thus estimated the terminal velocity to equal 0.612 feet per second.

We then compared our solution for terminal velocity by Euler's Method with that calculated directly from the differential equation, $\dfrac{dv}{dt} = g - cv$. When the velocity approaches terminal velocity, $\dfrac{dv}{dt}$ approaches zero, because at terminal velocity there is no acceleration. When we set $\dfrac{dv}{dt} = 0$, we got $0 = g - cv$, and $v = \dfrac{g}{c}$. Plugging in the values, $g = 32.2$, and $c = 52.6$, we arrived at the terminal velocity of 0.612 feet per second. This value agreed with that obtained by Euler's Method.

Since the raindrop reaches terminal velocity quite rapidly, we could, in estimating the time required for the raindrop to fall 3,000 feet, assume that the velocity is the constant terminal velocity the whole time. We calculated the required time by using the relationship, $t = d/r$, where $t$ represents time, $d$ represents distance, and $r$ represents speed. We found the time to equal 4,900 seconds. We noticed that this value varies considerably with that found using the model that did not account for air resistance. We also calculated the exact solution for terminal velocity using the formula given in the lab for velocity as a function of time:

$$V(t) = \frac{g}{c}(1 - e^{-et}).$$

We graphed this function together with the graph of values obtained by Euler's Method and found them to agree very closely (Fig. 1). The agreement increased as we added on more steps. When using the formula for $V(t)$, we evaluated the function for a large $t$. In

EXAMPLE 3 exhibit -- continued

this case, $t$ equaled 100, the same value we used for $k$ in finding the terminal velocity by Euler's Method. $V(100)$ equaled 0.612 ft/sec, the same value we obtained by the other methods.

For the second section of the lab, we once again approximated terminal velocity using the differential equation, itself, as well as Euler's Method. In this case, however, we used a model for large raindrops of diameter 0.05 inches. The differential equation for the velocity of the large drop, using $w$ for velocity, was

$$\frac{dw}{dt} = g - \alpha w^2, \text{ where } \alpha \text{ is a constant with a value of } 0.115.$$

The initial condition, as before, was that $w(0) = 0$. We obtained this equation in similar fashion as we acquired the others, except that in this case, the force of air resistance was proportional to the square of the velocity. This could be represented symbolically as $F_r = pw^2$, where $p$ is some constant. Newton's Second Law states that $m\frac{dw}{dt} = mg - pw^2$. Dividing through by $m$, we get $\frac{dw}{dt} = g - \frac{p}{m}w^2$. We can then let the expression $\frac{p}{m}$ equal some constant $\alpha$, and our equation becomes $\frac{dw}{dt} = g - \alpha w^2$. Setting $\frac{dw}{dt}$ equal to zero, as we did before, we found the terminal velocity to equal 16.7 feet per second. Assuming the velocity to be the constant terminal velocity the entire time, we calculated the time required for the large drop to fall to the ground from a height of 3,000 feet to be 180 seconds. Compared to the 4,900 seconds it took for the small raindrop, the larger raindrop falls much faster.

Lastly, we approximated the graph of the velocity function by using Euler's Method. Employing the same methods as before, we found the recursive formula to be the following:

$$w_{k=1} = w_k + (g - \alpha w_k^2)\alpha t$$

We then graphed the velocities for various times $t$ (Fig. 2). We observed the graph to find a leveling-off point as before, one which would give us an idea of the terminal velocity. We then selected a large $k$ – again, 100– beyond the observed leveling-off point and had MathCAD evaluate $w_{100}$. This value turned out to be 16.7 feet per second, which agreed with the value obtained from the differential equation. In addition, compared to the smaller drop, the larger drop took between 10 and 15 times as long to reach its terminal velocity.

We concluded that if one chooses a large $k$, and thus a small $\delta t$, Euler's Method is effective at finding the solutions to initial value problems. The values we obtained using Euler's Method consistently agreed with those acquired using the differential equations and, in the small-drop problem, with that obtained from the exact, original function. In this case, the numerical method was just as effective as the calculus method in providing the desired solutions.

A note on the relation between Examples 2 and 3

There are many parallels between the Space Probe project and the Falling Bodies lab. The mathematics and the physics of the examples are quite similar. Both use technology to support graphing and computation, and both illustrate ways in which important assessment activities are intimately interwoven with instruction. However, in terms of the structure of the activity, there are important, albeit subtle, differences, reflecting the difference between a project used in a course whose primary mode of instruction is class-based and a structured laboratory that serves as the primary mode of instruction. In the former, the overall modeling activity was carefully structured and supported, but the building of the particular mathematical model was not: the students' primary job was to build, apply, document, and report on the model they built, including a critique of their model and the choices underlying it. In the latter, the building of the particular mathematical models was highly structured, and the ostensible purpose was learning the particular mathematical models and their features rather than learning the wider skills of modeling (a goal that is addressed in other parts of the course). It is worth noting that in both cases the underlying values come through quite clearly. The importance of using technology, of writing, and of collaboration are common to both, and they are hard to miss. In the case of the falling bodies example, there is also a shift from the standard calculus course in "what counts" mathematically. In the standard courses, the purely symbolic model is the goal; typically the situation provides the context for moving toward that goal. In this lab, however, the recursive model using Euler's Method is of equal status with the purely symbolic model.

**Example 4: Periodic Functions**

Students encounter the periodic functions lab about half way through the second semester of the course that also provided Example 3. When they work the lab, the students have completed work on symbolic evaluation of definite integrals using hand computations, a table of integrals, and the computer algebra system Derive. They have seen applications of the definite integral to the computation of areas and centers of mass and will soon look at the calculation of arc length.

Student work in this example is not, as in the preceding two examples, connected directly to a real-world problem. Rather, the project itself has the students discover the formulas for the Fourier coefficients of a periodic function. The connection with synthesis and analysis of periodic functions takes place in the classroom.

An important feature of this project is the use of Derive to enable students to make and test conjectures. The students are asked to examine the graphs of functions such as $\sin t \cos 2t$, $\sin t \sin 2t$, and $\cos t \cos 2t$, and conjecture what the value of the integral of such functions should be over the interval $[-\pi, \pi]$. The students test and revise their conjectures by having Derive calculate additional integrals. The process is repeated for integrals of functions of the form $\sin^2 kt$ and $\cos^2 kt$ over the same intervals.

Then, in a MathCAD worksheet, the students must use these conjectures to determine the coefficients of a trigonometric polynomial – a polynomial whose formula is unknown, but whose values may be obtained for graphing and numerical integration. The students experience considerable satisfaction when the graph of their proposed function matches the graph of the unknown trigonometric polynomial.

The role of the computer algebra system in this laboratory is central, as it is in the course as a whole. Without such facilities at their disposal, the students would focus on the computation of the integrals as the central task. With them, the students can graph and calculate numerous integrals; they focus on the understanding of why these integrals have the values they do and what they are good for.

The two student reports that are reproduced following the description of the lab exercise illustrate the fun that students can have with this project (not to be underestimated – how many students engage in such flights of fancy with conventional homework assignments?), and record the results of an additional investigation undertaken by a student in response to the lab. (Again, how frequently does this occur in general? And, would students consider undertaking such an investigation if they did not have a high comfort level with computational tools?) As such, both the assignment and the responses to it allow for inferences about the goals of the course, and how some students progressed toward them.

Student reports in general also bring to light student conceptions that instructors may not expect at first. A typical first conjecture is, "The integral of the product of any two sine and cosine functions over the interval $[-\pi, \pi]$ is 0." The team's second conjecture often is, "The integral of the square of any sine or cosine function over the interval $[-\pi, \pi]$ is $\pi$." When such a team is asked if the two conjectures might contradict each other, a common reply is, "No, one conjecture is about products, and the other is about squares." This response provides food for thought, for both instructors and students.

It is important to note that in the case of examples such as the ones discussed here – projects and laboratory reports – assessment and instruction are deeply intertwined. That is, providing students useful feedback on their current performance is a major component of the instruction itself; it is not merely an end-of-instruction procedure for assigning grades.

EXAMPLE 4 exhibit

---

# Instructions for Lab 22: Periodic Functions

## Purpose.

Many functions may be described as combinations of other simpler functions. Later in the semester we will consider (in Labs 24 and 25) representations of functions by combinations of power functions (i.e., by polynomials). Here we consider representations by combinations of sines and cosines (so-called "trigonometric polynomials"). In particular, we will investigate how integration may be used to find a representation by sines and cosines of a complicated periodic function.

## Preparation.

Before coming to lab, read pages 1 through 6 of Chapter 10, and do Exercises 1 through 7. Read the rest of these instructions, and bring the text, exercises, and instructions to lab.

## Project.

1. Bring up *Derive*, and declare $t$ to be a variable with an interval range $-\pi \leq t \leq \pi$. Graph the function $\sin 2t \cos t$; use the graph to *guess* the value of the integral $\int_{-\pi}^{\pi} \sin 2t \cos t \, dt$. Enter your guess here: _____ Use *Derive* to calculate the value of the definite integral: _____

Calculate the following integrals with *Derive*:

$$\int_{-\pi}^{\pi} \sin t \sin 2t \, dt = \text{_____} \quad ; \quad \int_{-\pi}^{\pi} \sin t \cos 2t \, dt = \text{_____}$$

On the basis of the three calculations above, make a conjecture about integrals of products of sine and cosine functions over the interval $-\pi \leq t \leq \pi$. Write your conjecture here:

EXAMPLE 4 exhibit -- continued

Calculate two more integrals to check your conjecture; choose integrands that are different from the ones already integrated, but that should illustrate your conjecture:

$$\int_{-\pi}^{\pi} \underline{\hspace{3cm}} \, dt = \underline{\hspace{3cm}}$$

and

$$\int_{-\pi}^{\pi} \underline{\hspace{3cm}} \, dt = \underline{\hspace{3cm}}$$

If these examples do not support your conjecture, make a new conjecture and test with new examples. Continue this process of conjecture and testing until you are sure your conjecture is correct. Write your conclusion here.

2. Use *Derive* to calculate the following integrals:

$$\int_{-\pi}^{\pi} \sin^2 t \, dt = \underline{\hspace{3cm}} \qquad \int_{-\pi}^{\pi} \sin^2 2t \, dt = \underline{\hspace{3cm}}$$

$$\int_{-\pi}^{\pi} \cos^2 t \, dt = \underline{\hspace{3cm}} \qquad \int_{-\pi}^{\pi} \cos^2 2t \, dt = \underline{\hspace{3cm}}$$

Make a conjecture about integrals of squares of sine and cosine functions, and enter it here:

## EXAMPLE 4 exhibit -- continued

Check your conjecture by evaluating two more integrals:

$$\int_{-\pi}^{\pi} \underline{\hspace{4cm}} \, dt = \underline{\hspace{4cm}}$$

and

$$\int_{-\pi}^{\pi} \underline{\hspace{4cm}} \, dt = \underline{\hspace{4cm}}$$

If these examples do not support your conjecture, make a new conjecture and test with new examples. Continue this process of conjecture and testing until you are sure your conjecture is correct. Write your conclusion here.

Does your conjecture about integrals of *squares* (a kind of product) contradict your conjecture about *products*? Why or why not?

3. In the next part of this lab project, you will use a *MathCAD* worksheet. Quit *Derive*, bring up *MathCAD*, and load LAB22. In the places indicated on the worksheet, enter the month of birth (1 for January up to 12 for December) and the day of the month of birth (1 to 31) of the team member currently at the keyboard. These values determine a function of the form

$$f(t) = b_0 + b_1 \cos t + a_1 \sin t + a_2 \sin 2t \, .$$

The definition of this function is hidden from your view, but *MathCAD* can calculate its values whenever you ask it to. Graph your function in the box provided.

EXAMPLE 4 exhibit -- continued

Your task is to use what you learned from Exercises 5, 6, and 7 in Chapter 10, plus parts 1 and 2 of this Project, plus *MathCAD*'s integration routine, to find the values of the constants $a_1$, $a_2$, $b_0$, and $b_1$. Use the space below the graph box for your calculations. As you find each constant, enter it in the appropriate place at the top of the worksheet. When you redraw the graph, the function $g(t)$ (graphed with + signs) will record your progress in approximating $f(t)$; when the two match, you are done. Print the final worksheet showing the matching functions and the calculations used to find the coefficients. (Make two copies; each member of the team needs one.)

### Report.

The work done in the lab is a joint effort of the team, but (for this lab only) each student must write a separate expository report on the questions raised. Explain the conjectures you made, how and why you changed them, and their final status. Explain in your own words the role of the integral calculations as a tool for finding unknown coefficients in a trigonometric polynomial. You do not need to include solutions to the exercises in the text; you may use those results as needed, provided you cite the appropriate exercise for each such result. Include a copy of your (joint) *MathCAD* worksheet with your report.

EXAMPLE 4 exhibit -- continued

### Lab 22: Periodic Functions

**Enter the month of your birth as m (1 through 12)** $\qquad m \equiv 4$

**Enter the day of your birth as d (1 through 31)** $\qquad d \equiv 6$

**The coefficients of your estimation of the formula for f are set initially to 0.**

$$b_0 := \left[\frac{1}{2 \cdot \pi}\right] \cdot \int_{-\pi}^{\pi} f(t)\, dt \qquad\qquad b_1 := \frac{1}{\pi} \cdot \int_{-\pi}^{\pi} f(t) \cdot \cos(t)\, dt$$

$$a_1 := \frac{1}{\pi} \cdot \int_{-\pi}^{\pi} f(t) \cdot \sin(t)\, dt \qquad\qquad a_2 := \frac{1}{\pi} \cdot \int_{-\pi}^{\pi} f(t) \cdot \sin(2 \cdot t)\, dt$$

$$g(t) := b_0 + b_1 \cdot \cos(t) + a_1 \cdot \sin(t) + a_2 \cdot \sin(2 \cdot t)$$

**Graph the function  f ( t )  and the current approximation,  g ( t ) :**

$$t: -3 \cdot \pi, -3 \cdot \pi + .1 \ \ldots\ 3 \cdot \pi$$

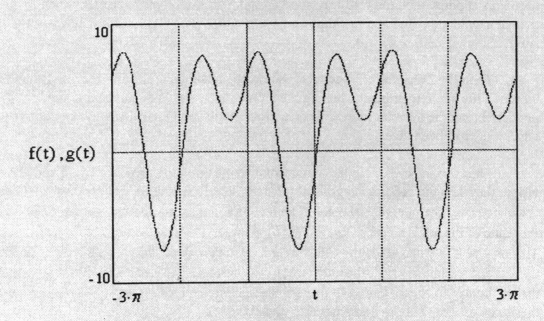

**Lab Sheet for Lab 22**

EXAMPLE 4 exhibit -- continued

# The Gospel According to

# ISAAC

## transcribed and interpreted by the prophets

### Doug and Dave

In the beginning, there was periodic function. And the function was formless and without visible pattern. And darkness was over the face of those who sought the form of periodic function. And Isaac said, "Let there be Calculus!" And there was Calculus, and the mathematics world saw that it was good (though somewhat abstract). Then came forth Leibniz, claiming, "Behold, I have also created a world of Calculus, and desire that it should be used like unto that of Isaac." But the mathematics world answered him, saying, "Great Leibniz, it is true you have created a world like unto Isaac's. But though thy name is therefore to be revered, it is also impossible to spell." And so Isaac was supreme in the Skies of Mathematics, as he is even today (despite the claims of the Heretical Leibnizian prophets -- beware their treacherous false teachings!)."

And when the Great Teacher Dr.** saw that his followers had no knowledge of periodic functions, he opened his mouth and spoke to them saying, "Know thou this lab which I have given unto you. Follow thou the instructions and sanctify thyself in the exercises which I will show you, that ye may be prepared and thy knowledge strong in the day when you shall come unto me in lab."

Then the Disciples of Isaac went their way, chanting "Calculus Regnum," and buried the words of the lab in their hearts, fearing the wrath which would befall them in the lab. They pondered long the introductory remarks, and with skill and care calculated the definite integrals which they had been given.

Then, on the eve of the lab, a revelation came upon the prophets Doug and Dave. In the vision, they saw the definite integrals of the periodic trig functions sine and cosine. They saw them calculated over the interval between $-\pi$ and $\pi$. A voice called to them, saying, "Lo! Whatsoever the value of the angle measure, the value of the integral between $-\pi$ and $\pi$ for sine and cosine is 0. This may ye know and teach thy children, and thy children's children." And the vision revealed that, over the same interval, the integral for the constant 1 was $2\pi$. And they awoke and went to the lab, unable to speak, having in their minds the knowledge that great and mighty things, which they knewest not, were to be revealed to them there.

And it came to pass that in the lab, the prophets called upon the great DERIVE, which knowest all things, saying "Calculus Eternae." And they asked DERIVE to draw them a picture of the periodic function,

$$\sin(2t)\cos(t),$$

## EXAMPLE 4 exhibit -- continued

that they might guess the integral of the function between $-\pi$ and $\pi$. And when they had seen it, they saw the area beneath the curve of the graph on the positive side seemed equal in value to the area above the curve on the negative side. Therefore they foretold that DERIVE would show the value of the integral to be 0. And lo! When DERIVE had calculated the integral of the function, it was 0! And the prophets saw that they had guessed right, and were glad, saying, "Gratis Calculae!".

The Lab Commandments declared that they should calculate yet another integral over the interval $-\pi$ to $\pi$, this time the function,

$$\sin(t)\sin(2t).$$

And when they had done it, they saw what DERIVE was telling them, that the value of this integral was also 0. The prophets were commanded to write what they had seen, and they recorded the revelation: "Behold, the value of the integral from $-\pi$ to $\pi$ of any periodic trig function, when multiplied either by a constant or by another periodic function, shall be zero, unless the function is the square of some periodic function (like $\sin^2(t)$)."

And it came to pass that Lab commandments asked that the prophets calculate still more integrals, which they should create of their own mind, so that they might check their conjecture. The prophets chose the two functions,

$$\frac{1}{2}\sin(t)\cos(2t)$$

and

$$\frac{1}{2}\sin(2t)\cos(t).$$

The prophets saw that they could test more functions, but they had to move on.

And then they came to a new portion of the lab which they had not yet seen: the third Commandment told them to implore the mighty DERIVE to calculate the integrals of the squares of some periodic functions, again over the integral $-\pi$ to $\pi$:

$$\sin^2(t)$$

$$\sin^2(2t)$$

$$\cos^2(t).$$

And when it was done, they found that DERIVE had shown that the value of all these definite integrals was the Great Constant, $\pi$. It then commanded that they should again write the revelation: "Behold, the value of the integral of the squares of the periodic trig function Sine over the interval $-\pi$ to $\pi$ shall be $\pi$ for ever and ever, amen. And so shall it be for Cosine, when the variable is multiplied by some constant greater than 1. So let it be forever." This law was put to

EXAMPLE 4 exhibit -- continued

the test using the function $\sin^2\left(\frac{1}{2}t\right)$ and $\cos^2(3t)$, which did equal the Great $\pi$ in the Sky; and for

$\cos^2\left(\frac{1}{2}t\right)$, which did not.

*[Instructor's Note on paper:* $\displaystyle\int_{-\pi}^{\pi} \cos^2\left(\frac{1}{2}t\right)$ *is* $\pi$. *However, if you replace* $\frac{1}{2}t$ *by* $\frac{1}{3}t$,

*you have a problem. In any event,* $\displaystyle\int_{\cos^2}^{\pi} ntdt = \int_{-\pi}^{\pi} \sin^2 ntdt = \pi$ *for all non-zero integers n.]*

   Then the time had come that the prophets would be tested according to that which had been revealed to them regarding the periodic function. For lo, there lay lurking in the realm of darkness a function of unknown coefficients. The prophets paid their homage to the great **DERIVE** and travailed the way to the Temple of MathCAD, where they found awaiting them the spirit of a function of the form

$$f(t) = b_0 + b_1 \cos(t) + a_1 \sin(t) + a_2 \sin(2t).$$

The prophets were left without knowledge of the values of the $a$'s and $b$'s of the function —yea, it was their mission to find those values using the knowledge bequeathed to them earlier.

   They began by imploring MathCAD to draw them a picture of the function, since MathCAD had knowledge of the functions which they knew not. They began their search for the values of the constants by seeking to show that

$$2\pi b_0 = \int_{-\pi}^{\pi} f(t)\, dt.$$

The prophets plugged in what they had:

$$\int_{-\pi}^{\pi} b_0 + b_1\cos(t) + a_1 \sin(t) + a_2 \sin(2t)\, dt =$$

$$\int_{-\pi}^{\pi} b_0 dt + \int_{-\pi}^{\pi} b_1 \cos(t)\, dt + \int_{-\pi}^{\pi} a_1 \sin(t)\, dt + \int_{-\pi}^{\pi} a_2 \sin(2t)\, dt$$

From what they had learned earlier by calculating integrals, the last three terms, which were the integrals of periodic trig functions, equaled 0, and therefore dropped out of the equation:

$$\int_{-\pi}^{\pi} f(t)\, dt = \int_{-\pi}^{\pi} b_0\, dt.$$

They also knew from the First Commandment that the value of the remaining integral equaled $2\pi$ times the constant, $b_0$. Thus,

$$\int_{-\pi}^{\pi} f(t)\, dt = 2\pi b_0.$$

## EXAMPLE 4 exhibit -- continued

The prophets then implored MathCAD to reveal the value of the integral of $f(t)$ from $-\pi$ to $\pi$: which MathCAD knew. That value was 19.918, and by setting that number equal to $2\pi b_0$, they could solve for $b_0$.

The next step in finding the value of the constants was to multiply both sides of the equation for $f(t)$ by some trig function such that all but one of the terms of the equation would drop out. The prophets deemed it most prudent to first solve for $b_1$. They saw that, by multiplying $f(t)$ by $\cos(t)$, the other three terms would drop out:

$$\int_{-\pi}^{\pi} \cos(t) f(t)\, dt = \int_{-\pi}^{\pi} \cos(t)\, b_0\, dt + \int_{-\pi}^{\pi} b_1 \cos^2(t)\, dt + \int_{-\pi}^{\pi} a_1 \cos(t)\sin(t)\, dt + \int_{-\pi}^{\pi} a_2 \cos(t)\sin(2t)\, dt.$$

Thus,

$$\int_{-\pi}^{\pi} \cos(t) f(t)\, dt = \int_{-\pi}^{\pi} b_1 \cos^2(t)\, dt,$$

$$\int_{-\pi}^{\pi} \cos(t) f(t)\, dt = b_1 \pi.$$

Again, MathCAD was consulted for the value of the left side of this equation, which was revealed to be 17.687. Solving for $b_1$ (by dividing both sides by $\pi$), they found its value to be 5.63.

The prophets had hit upon a pattern, and continued to multiply by trig functions which would make three of the four terms equal to 0, then use MathCAD to calculate the corresponding integral of $f(t)$, and solve algebraically for the remaining constants. For the $a_1$ term, they multiplied by $\sin(t)$; the integral of $f(t) \sin(t)$ was -13.603 and $a_1 = -4.33$. For the last term, $a_2$, they multiplied by $\sin(2t)$; the integral was 18.473, and $a_2 = 5.88$.

Finally, the prophets told MathCAD the function which they had discovered, with the constants which they had prophesied. They implored MathCAD to draw a picture of this new function,

$$g(t) = 3.17 + 5.63\cos(t) - 4.33\sin(t) + 5.88\sin(2t)$$

over the picture of the old function, $f(t)$. When this was done, no difference could be found between the graphs. With exceeding joy the prophets gave thanks to MathCAD and to DERIVE, saying, "Hark! We have learned something... for we knewest not the constants of our hidden function. And lo! The wondrous pleasure of using our knowledge and the revelations of DERIVE to find that which we sought!" And the prophets Dave and Doug went their ways, chanting, "Calculus Vitae. Regnus sum omnum!" (and various and sundry words of praise and thanksgiving to Isaac). And they slept deeply and peacefully, knowing that what they had seen and done would be pleasing in the eyes of the Great Teacher, Dr. **.

EXAMPLE 4 exhibit -- continued

# Approximating Integrals of Powers of Sine and Cosine

As we mentioned above, the value of the area under the curve described by an even power of a sine or cosine function is $\pi$ times some fractional coefficient. In an independent exploration, I sought some way to calculate the coefficients of $\pi$ without actually evaluating the integral of the trigonometric function (over the interval from $-\pi$ to $\pi$). I was unable to discern any definite pattern among the coefficients, but I was able to approximate the coefficients as a function of the power of the trigonometric function.

With the help of *Derive* I compiled the following table of coefficients of $\pi$ as a function of the powers of the trigonometric functions.

| Power | Coefficient |
|-------|-------------|
| 2 | 1 |
| 4 | $\frac{3}{4}$=0.75 |
| 6 | $\frac{5}{8}$=0.675 |
| 8 | $\frac{35}{64}$=0.546875 |
| 10 | $\frac{63}{128}$=0.4921875 |
| 12 | $\frac{231}{512}$=0.451171875 |
| 14 | $\frac{429}{1024}$=0.418945312 |
| 16 | $\frac{6432}{10384}$=0.39276123 |
| 18 | $\frac{12155}{32708}$=0.370941162 |

The graph of coefficients versus the power of the trig function seemed to have some sort of exponential or power relationship. A log–log plot changed the graph to a straight line of the form $y = mx + b$, indicating the power relationship. Using *MathCAD* I solved for the slope of the line and its intercept and obtained an equation for approximating the coefficient of $\pi$ as a function of the power of the trig function.

$$\text{coefficient} = c^{0.373} \text{x power}^{-0.474}$$

Note that this is only an approximation; however, the differences between the approximations and the actual values for the coefficients is small. More importantly, this approximation saves time in calculating the integral of an even–powered trigonometric function when the power is large.

**Example 5: A sequence of instructional examples related to the convergence of a sequence or a series of functions**

The sequence of activities described here takes place in the $C^4L$ project at Purdue. The project makes extensive use of computer programming, in the programming language ISETL, whose syntax is very close to standard mathematical notation. As described below, some of the instruction in the course is contingent upon the assessment of student work: decisions to present material, or to pull things together, are based on the instructor's determination that students have begun to construct certain understandings. There is extensive student group work and discussion in the course and a minimal amount of lecturing. In this context, assessment is an integral part of instruction and is grounded in research on students' understanding of the function concept.

The activities are concerned with the convergence of a sequence or series of functions – a topic of significant difficulty. Part of the issue is to understand what an expression such as

$$f(x) = \sum_{n=0}^{\infty} f_n(x)$$

means from a student's perspective, and how to foster an understanding consistent with the mathematician's. In a series of labs, students work in teams to write computer programs that represent functions defined by specific series of functions. They learn to write programs that (a) accept a number $x$ and return a function which represents a sequence of numbers in that, given a positive integer $n$, the program returns a number $f_n(x)$; and (b) perform the summation over a collection of the $f_n(x)$. When the students have completed a unit on series of numbers, they are given tasks such as the following.

1. Consider the following list of infinite series. Notice that in each series, the general term is not a number, but an expression that represents a function. Thus, for each value of $x$, you have a different series.

(a) $\qquad f(x) = \sum_{n=0}^{\infty} x^n$

(b) $\qquad f(x) = \sum_{n=0}^{\infty} nx^{n-1}$

(c) $\qquad f(x) = \sum_{n=0}^{\infty} \frac{x^{n+1}}{n+1}$

For each of these series, write a program that takes as parameter a value $x$ and returns a function that represents the corresponding sequence of partial sums.

2. In the series of Activity 1(a), investigate convergence for the values $x = -1.5, -1, -0.5, 0.5, 1, 1.5$.

3. For the series of Activity 1(a), write a program **ser** that represents a function whose domain is a set of real numbers. Your program should accept a value $x$ as parameter and construct a function that represents the sequence of general terms for the given value of $x$. Then apply your program **PS** (partial sums) from the unit on sequences and series of numbers to obtain a function that represents the sequence of partial sums. Finally, your program should return the value of the 50th partial sum.

Use your table facility to generate a list of 20 values of your function, including the 6 values from Activity 2 above.

This is followed by activities in which the students are asked to repeat the task for the other two series, to experiment with replacing the number of terms (50) with smaller and larger values, and to investigate term-by-term sums, products, derivatives, and integrals of series.

One possible solution for the first three activities is as follows. Notice the similarity between the programming syntax and standard mathematical notation (the main difference being the use of %+ in place of $\sum$ ).

```
PS := func(s);  $ s is a sequence
    return  func(n) ;
        return  %+[s(i) : i in [1..n]];
    end;
  end;
ser := func(x);
    return  func(n) ;
        return x**n;
    end;
  end;
F := |x -> PS(ser(x))(50)|;
```

After the students have completed these tasks in the computer laboratory, they are asked to work in their teams on each of the following tasks with paper and pencil. The work is discussed in class.

1. Consider the series

$$\sum_{n=0}^{\infty} \frac{x^n}{\sqrt{n+1}}$$

(a) Find a value of $x$ for which this series converges. Others?

(b) Find a value of $x$ for which this series diverges. Others?

(c) Can you describe a general method for determining convergence?

(d) Find all values of $x$ for which this series converges.

(e) What happens if $x$ is replaced by $x+1$?

2. In the series of the previous task, is there a function being defined? What is its domain? Is there a formula for it?

At this point in the class, if the instructor thinks that students have made significant progress in constructing the ideas, there is the option of providing a brief lecture giving careful definitions of power series, base point, coefficients, radius of convergence, and interval of convergence. The students then work on similar tasks with the other two series and investigate combinations of series and the representation of functions, their derivatives, and their integrals by series. With suitable modifications of the function F in the program above (e.g., replacing the value 50 by other numbers, thus exploring larger and smaller partial sums; replacing the expression x**n by other expressions), the student has constructed empirical tools for studying issues such as the convergence of series of functions, approximation of functions, representation of functions by series, and term-by-term differentiation and integration of series.

Discussion

This set of activities tends to bring out into the open aspects of students' process and object conceptions of function (cf. Section 6.1) related to sequences and operations on them. The syntax of the programs PS,

ser, and F (see the example above) presents students with great difficulty, and it takes a long time before the students can get the programs to run on a computer. At this point, the students complain about the complexities of the programming language. However, experience suggests that the difficulties are conceptual; that, for example, the difficulty students have in writing a line such as

```
return func(n);
```

even after seeing it in other examples, is a consequence of their inability to move from thinking of a function as a process to conceptualizing it as an object. Once that difficulty is overcome, students are able to write functions that return functions.

More generally, the activities discussed here reflect a particular approach to instruction that is based on an analysis of student understanding and that has assessment activities, keyed to that analysis, embedded within the instruction.

**Example 6: A 2 1/2 hour in-class final examination**

The Calculus 1 exam that follows comes from Project CALC. Project CALC has weekly computer labs that place an emphasis on applied problems and on student writing. It is worth reflecting on the balance of tasks in this exam and what it reflects about the emphases of the course; also about the specific instructions students are given for taking the exam. One sees that calculations (Part I of the exam) do play a role in the course, but that the role is circumscribed: students are told to spend 20 percent of the exam time on Part I. Due to time limits, the "concepts" problems (Part II of the exam) are somewhat constrained; they bear some resemblance to standard word problems found on typical calculus exams. However, the instructions for working the problems – "If you find that a calculation leads to an unreasonable answer, you will get more credit for identifying the apparent problem and saying why the answer is unreasonable, less credit for not noticing or pretending there is no problem. Of course, you will get still more credit if you find out what went wrong in the calculation and fix it" – are decidedly nonstandard, and clearly place the emphasis on understanding and explaining rather than on producing one right answer. The essay questions (Part III of the exam) again stress writing and explanations: "Explain your answer in a manner that someone unfamiliar with calculus can understand. Write your answer in complete sentences that form a coherent paragraph or paragraphs." Note that Problem 1 of Part III calls for the qualitative analysis of graphs, and for making a prose argument about the properties discovered. As such, this problem is a more advanced example of the kinds of problems on multiple representations and qualitative analysis discussed in Section 6.1. The applied, tool-based focus of the course is reinforced by an essay question on approximation, and the fact that students had access to various materials, including calculators, during the exam.

EXAMPLE 6 exhibit

# Final Examination, Calculus I
### Fall Semester, 1991

**Instructions:**

*This exam has three parts: Calculations, Concepts, and Essay Questions. We suggest the following time allocations (which also reflect the relative importance of each section in our grading): Calculations, 30 minutes; Concepts, one hour; Essay, one hour. If you have time remaining, use it for checking, refinement, and correction.*

*If you find that a calculation leads to an unreasonable answer, you will get more credit for identifying the apparent problem and saying why the answer is unreasonable, less credit for not noticing or for pretending there is no problem. Of course, you will get still more credit if you find what went wrong in the calculation and fix it. You may use your text, project, and lab materials, your tables, your calculator, and your notes.*

**Part I: Calculations.  Show your work.**

1.  Find the derivatives of the following functions:

(a)     $f(x) = 3\,e^{-4x}$

(b)     $f(x) = \sin(x^2)$

(c)     $f(x) = \dfrac{x^2}{1 + 2x}$

2.  Find (approximately) a number x which solves the equation $x = 5\,\ln x$.

**Part II: Concepts.  Write your answers in complete, connected sentences.**
                 *(Note: Individual instructors selected problems for this section from among the following. No one instructor used all of these problems.)*

1.  A patient's "reaction" $R(x)$ to a drug dose of size $x$ is given by a formula of the form
$$R(x) \;=\; A\,x^2\,(B - x),$$
where $A$ and $B$ are positive constants.  The "sensitivity" of the patient's body to a dose of size $x$ is defined to be $R'(x)$.

   (a)    What do you think the domain of $x$ is?  What is the physical meaning of the constant $B$?
          What is the physical meaning of the constant $A$?
   (b)    For what value of $x$ is $R$ a maximum?
   (c)    What is the maximum value of $R$?
   (d)    For what value of $x$ is the sensitivity a maximum?
   (e)    Why is it called "sensitivity"?

2.  A truck traveling on a flat interstate highway at a constant rate of 50 MPH gets 4 miles to the gallon. Fuel costs $1.15 per gallon. For each mile per hour increase in speed, the truck loses a tenth of a mile per gallon in its mileage. Drivers get $27.50 per hour in wages, and fixed costs for running the truck amount to $12.33 per hour. What constant speed should a dispatcher require on a straight run through 260 miles of Kansas interstate to minimize the total cost of operating the truck?

3.  At a certain instant, just before lifting off, a plane is traveling down the runway at 285 kilometers per hour. The pilot suddenly realizes something is wrong and aborts the takeoff by cutting power and applying the brakes. Assume that the effect of this action is a deceleration proportional to time $t$. After 28 seconds the plane comes to a stop. How far does the plane travel after the takeoff is aborted?

## EXAMPLE 6 exhibit -- continued

4. For what point(s) on the curve $y = x^2$ do(es) the tangent line to $y = x^2$ go through the point (3,5).

5. A spring of mass 4 kilograms and spring constant 9 kg/sec$^2$ is pulled in the direction away from a wall one meter beyond the equilibrium point. It is released from this point with an initial velocity (towards the wall) of 1.5 meters/sec.

   (a)  Write down the differential equation and initial conditions satisfied by the function, $x(t)$, which gives the displacement from equilibrium as a function of time .
   (b)  Solve to find $x(t)$.
   (c)  At what times will the mass be closest to the wall?

**Part III. Answer each question in the form of a short essay.**

1. Below are three graphs. One is the graph of $f(x)$, another is $f'(x)$ and another is $f''(x)$. Identify which is which and carefully explain your reasons in a paragraph or two.

2. Approximate, to one decimal place, the instantaneous rate of change of the function

$$f(t) = 5^{\sqrt{\ln t}} \text{ at } t = 2.$$

Explain your answer in a manner that someone unfamiliar with calculus can understand. Write your answer in complete sentences that form a coherent paragraph or paragraphs.

### Example 7: A 2-Day Take-Home Examination

This final example is of a 2-day open-book, take-home Calculus II examination from the Calculus in Context project of the Five Colleges Consortium. The project places a major emphasis on applications and the use of technology.

A quick scan of the exam with regard to the assessment framework described in Section 3.1 shows both the emphases of this exam and the scope of coverage. We list the categories about which it is possible to make inferences from the exam.

7 A, G. Philosophical and pedagogical goals, as reflected in the constraints for taking the exam.

Some major goals of instruction, and assessment for this project, are made clear in the conditions laid down for taking the exam: For the purposes of this exam, any resources except other people are considered to be fair game. For example, it is entirely within bounds to find the topic in a book and (as is most likely to be necessary) adapt what is in the book to solve the task. Indeed, pointers to the literature are prominent in the examination (e.g., the Richardson Arms Race Model is named in the problem statement), should the student wish to follow them. Access to technology is assumed, but the student is given no pointers: it is one of many tools, and if the students are to use it, they have to figure out how and where. These conditions are very un-school-like, and (with the exception of the constraint that work must be done by individuals only) quite like the demands of the workplace. The job is to get the job done—and knowing how to use the resources in the environment is part of getting the job done. The emphasis of the applied problems is on the interpretation and application of the models, but explanation and communication also play a significant role.

7 B, C, D, E: Content, thinking processes, products, the pure/applied spectrum.

The mathematical content is clear. It is, in some sense, standard second-semester calculus material, although what students are asked to do with the material is non-standard. There are both pure and very applied problems, the latter calling for interpreting and extending models

rather than creating them. Where things differ even more markedly from the norm, and where the framework described in Section 3 provides a means for documenting the breadth of what is being assessed, is with regard to thinking processes and products. Students working this examination are called upon to explore, experiment, and investigate; to do analysis, interpretation, conjecturing, and justifying—and more. They must interpret models, and provide coherent written explanations of their work. Note the comparison with Example 1 (and by extension, an exam on which Example 1 would be a typical problem). The comparison is revealing both in terms of what it says about the course and in terms of what students' work would reveal about what the students have learned.

EXAMPLE 7 exhibit

---

Calculus II

Second take-home test
Time limit: two days

Answer all questions. This is an "open-book" test; you may consult freely your notes, homework, text, and any other books you wish; you may use a calculator or a computer, and any programs available on a computer. However, you must not give or receive help, in any form, from anyone else. Make your responses brief but complete, and explain your reasoning. Do not leave these questions or your answers where anyone else can see them. Return your paper to the Science Center office after two days, but in no case later than 4:30 pm Friday, 5 April.

1.  The value of the integral

$$J = \int_1^\infty x^p \, dx$$

depends on the value of $p$. For example, if $p = -2$ then $J = 1$, while if $p = 1$ then $J = \infty$.
   a)  Determine all values of $p$ for which $J = \infty$.
   b)  Determine all values of $p$ for which $J$ is finite, and determine the value of $J$.

2.  a)  Obtain a formula for the Fourier sine transform of $f(x) = x$ on the interval $0 \le x \le 1$. That is, determine

$$F_s(w) = \int_0^1 x \sin(2\pi wx) \, dx.$$

   b)  Sketch the graph of $y = F_s(w)$ on the interval $0 \le w \le 5$. At what point $w$ on this interval does $F_s(w)$ attain its maximum? What is the maximum?

3.  *Predator-prey interactions with harvesting.* The basic models of predator-prey interactions take no outside environmental factors into account. This question adds one such factor—the effects of certain human intervention, called harvesting with equal effort.

   Consider a population of insects, such as moths, that damage an agricultural crop. Suppose the moths are kept under control naturally by a predator—in this case, spiders. But suppose the crop is sprayed repeatedly with an insecticide like DDT to reduce the moth population even more.

EXAMPLE 7 exhibit -- continued

Does the strategy work? The problem is that both spiders and moths die from the poison, so the DDT may do more harm than good.

Here is another example, connected with Volterra's original studies of fish catches in the Adriatic Sea. One species preys upon the other, and both are caught in nets. Continual fishing reduces the breeding population of each species. How does this affect the sizes of the two populations?

Proceed in the following way. Assume that the original predator-prey interaction is modeled by the standard equations in which the prey population grows logistically in the absence of predators:

$$x' = ax\left(1 - \frac{x}{K}\right) - bxy$$

$$y' = cxy - ey.$$

Here $x$ and $y$ are the sizes of the prey and the predator populations, respectively. Spraying or fishing (or more generally, "harvesting") removes a fraction of the breeding populations of both species. In the absence of any more precise information, we assume it is the *same* fraction—$h$. (This is what we mean by harvesting with "equal effort.") Thus we must decrease $x'$ by $hx$ and $y'$ by $hy$. Making these modifications to the basic equations, we get

$$x' = ax\left(1 - \frac{x}{K}\right) - bxy - hx$$

$$y' = cxy - ey - hy.$$

Now carry out an analysis of the following concrete problem:

$$x' = .1x\left(1 - \frac{x}{2500}\right) - .005xy - hx$$

$$y' = .00004xy - .04y - hy.$$

a) Assume first that $h = 0$. That is, examine the model before any harvesting occurs. Make a sketch in the $(x:y)$ plane that shows where $x' = 0$ and where $y' = 0$. Find the equilibrium values of this system, and mark them clearly on your sketch.
b) Suppose we begin with $x = 2000$ and $y = 10$. What happens to $x$ and $y$ over time?
c) Now harvest the populations by setting $h = .02$. Sketch the *new* places where $x' = 0$ and $y' = 0$, and find the *new* equilibrium values of $x$ and $y$. How does harvesting change the equilibrium of the system?

EXAMPLE 7 exhibit -- continued

d) Suppose we again start with $x = 2000$ and $y = 10$ but have a harvesting effort of $h = .02$. What happens to $x$ and $y$ now?

e) The basic model has an equilibrium with both predators and prey present (that is, $x > 0$ and $y > 0$). What happens to this equilibrium when harvesting occurs? At the new equilibrium, are there more or fewer predators, and are there more or fewer prey?

f) Does the model suggest that fishing increases the proportion of predator species in relation to the prey, or is it the other way around?

g) Does the model support the use of DDT to reduce the population of moths? Explain your position clearly.

4. *The Richardson Arms Race Model.* Between the two World Wars, the British physicist, Lewis Fry Richardson, devised a simple model to describe the "arms races" carried on by various nations at various times. It concerns two aggressively hostile countries or alliances of countries; call them X and Y. Let $x(t)$ represent the level of hostile activity of X at any time $t$, and let $y(t)$ represent the same for Y. For example, take $x$ to be the annual armaments budget for X, in billions of dollars, and measure time $t$ in years. There are diverse political pressures in X, some tending to make $x$ increase, some tending to make it decrease. The same is true in Y. Richardson concentrates on three sources of pressure:

1. The larger $y$ is, the greater the pressure to increase $x$. [An example: in the past decade, the U.S. navy has grown in size, partly because the Soviet navy has grown.]

2. The larger $x$ is, the greater the pressure to reduce $x$. [An example: it is costly for the United States to maintain large garrisons around the world; some in Congress argue the money would be better spent supporting social programs at home.]

3. Citizens of X may have a grievance against Y, independent of the size of either's armaments. [An example: Until quite recently, communism was widely considered to be repugnant in the United States, and the Soviet Union was labeled "an evil empire."]

Richardson assumes that $x$ changes in response to each of these pressures, and he expresses the *rate* at which $x$ changes by the following differential equation:

$$x' = \underbrace{ay}_{\text{Y's arms}} - \underbrace{mx}_{\text{X's arms}} + \underbrace{g}_{\text{grievance}}$$

A similar equation describes how $y$ changes:

$$y' = bx - ny + h$$

The values of the coefficients $a$, $b$, $m$, $n$, $g$, and $h$ are to be determined by the circumstances of a particular case.

EXAMPLE 7 exhibit -- continued

Now consider two possibilities: in the first, the type 2 pressures are larger than the type 1 pressures for both X and Y. In other words, each prefers to spend its money on domestic needs rather than armaments. Here is a specific example:

$$x' = .1y - .3x + 4$$

$$y' = .2x - .5y + 3$$

For the second possibility, suppose that the pressures for X are reversed while those for Y are unchanged:

$$x' = .3y - .1x + 4$$

$$y' = .2x - .5y + 3$$

a) For each of the two specific possibilities, determine whether there are hostility levels $x$ and $y$ for which the various pressures exactly balance, so that $x$ and $y$ do not change over time. If so, what are those levels?

b) Suppose the current hostility levels are $x = 10$ and $y = 20$. For each of the two specific possibilities, what are the hostility levels after one year and after two years? What happens in the long run?

c) What is the essential difference between the two possibilities presented above? Explain *how* the outcomes are different, and explain *why* they are different.

# References

Ayers, T., Davis, G., Dubinsky, E., and Lewin, P. (1988). Computer experiences in learning composition of functions. *Journal for Research in Mathematics Education,* 19(3), 246-259.

Breidenbach, D., Dubinsky E., Hawks, J., and Nichols, D. (1991). Development of the process concept of function. *Educational Studies in Mathematics.,* pp. 247-285.

Burkhardt, H. (1981). *The Real World and Mathematics.* London: Blackie & Son.

Cohen, M., Gaughan, E., Knoebel, A., Kurtz, D., and Pengelley, D. (1992). *Student Research Projects in Calculus.* Washington, DC: Mathematical Association of America.

Douglas, R. (Ed.). (1987). *Toward a Lean and Lively Calculus.* Washington, DC: Mathematical Association of America.

Dubinsky, E., and Ralston, A. (1993). *Undergraduate Mathematics: Now and Then.* Unpublished manuscript.

Elson, J. (1990). The test that everyone fears: A major revision shakes up the all-important SATs. *Time,* November 12, pp. 93-94.

Grouws, D.A. (Ed.). (1992). *Handbook of Research on Mathematics Teaching and Learning.* New York: Macmillan Publishing Company.

Harnisch, D. (in press). Performance assessment in review: New directions for assessing student understanding. *International Journal of Educational Research.*

Harvard Calculus Project. (1992). *Calculus Preliminary Edition.* Cambridge, MA: Harvard University Department of Mathematics.

Janvier, C. (1987). (Ed.). *Problems of Representation in the Teaching and Learning of Mathematics.* Hillsdale, NJ: Erlbaum.

Karian, Z. (Ed.). (1992). *Symbolic Computation in Undergraduate Mathematics Education.* Washington, DC: Mathematical Association of America.

Leinbach, C., Hundhausen, J., Ostebee, A., Senechal, L., and Small, D. (1991). (Eds). *The Laboratory Approach to Teaching Calculus.* Washington, DC: Mathematical Association of America.

Madaus, G., West, M., Harmon, M., Lomax, R., and Viator, K. (1992). The influence of testing on teaching math and science in grades 4-12. Cited in *National Science Foundation News*, NSF PR 92-86, October 15, 1992. Information on the study is available from the Center for the Study of Testing, Evaluation, and Educational Policy, 323 Campion Hall, Boston College, Chestnut Hill, MA 02167.

Markovitz, Z., Eylon, B., and Bruckheimer, M. (1983). Functions - Linearity unconstrained. *Proceedings* of the 5th meeting of the International Group for Psychology and Mathematics Education.

Massachusetts Department of Education. (1990). Beyond paper and pencil: Massachusetts educational assessment program. Quincy, MA: Author.

Mathematical Sciences Education Board (1993). *Measuring Up: Prototypes for Mathematics Assessment.* Washington: National Academy Press.

Mathematical Sciences Education Board (1990). *Reshaping School Mathematics: A Philosophy and Framework for Curriculum.* Washington: National Academy Press.

Moschkovich, J., Schoenfeld, A. H., and Arcavi, A. A. (1993). Aspects of understanding: On multiple perspectives and representations of linear relations, and connections among them. In T. Romberg, E. Fennema, and T. Carpenter (Eds.), *Integrating Research on the Graphical Representation of Function*, pp. 69-100. Hillsdale, NJ: Erlbaum.

National Council of Teachers of Mathematics. (1989). *Curriculum and Evaluation Standards for School Mathematics.* Reston, VA: Author.

National Council of Teachers of Mathematics. (1992). *Handbook of Research on Mathematics Teaching and Learning* New York: Macmillan Publishing Company.

National Council of Teachers of Mathematics. (1991). *Professional Standards for Teaching Mathematics.* Reston, VA: Author.

National Research Council (1989). *Everybody Counts: A Report to the Nation on the Future of Mathematics Education.* Washington: National Academy Press.

Newman, A. (1983). *The Newman Language of Mathematics Kit.* Sydney: Harcourt Brace & Jovanovich.

Pandey, T. (1990). Power items and the alignment of curriculum and assessment. In G. Kulm (Ed.), *Assessing Higher Order Thinking in Mathematics.* Washington, DC: American Association for the Advancement of Science.

Resnick, Z. (1987). *Functions* (Textbook from The Mathematical Chapters Series for 9th grade). (In Hebrew). Israel: The Department of Science Teaching, The Weizmann Institute of Science.

Schoenfeld, A. H. (1992). Balanced Assessment for the Mathematics Curriculum. *A Framework for Balance.* Technical Report BA-1992-01. Berkeley, CA: School of Education, University of California.

Schoenfeld, A. H., Smith, J. P., and Arcavi, A. A. (1994). Learning. In R. Glaser (Ed.), *Advances in Instructional Psychology.* Vol. 4, pp. 55-174. Hillsdale, NJ: Erlbaum.

Schoenfeld, A. H. (Ed.). (1990). *A Source Book for College Mathematics Teaching.* Washington, DC: Mathematical Association of America.

Schwingendorf, K., Hawks, J., and Beineke, J. (1992). Horizontal and vertical growth of the students' conception of function. In G. Harel, and E. Dubinsky (Eds.), *The Concept of Function: Aspects of Epistemology and Pedagogy.* MAA Notes, pp. 133-149. Washington, DC: The Mathematical Association of America.

Sfard, A. (1991). On the dual nature of mathematical conceptions: Reflections on processes and objects as different sides of the same coin. *Educational Studies in Mathematics, 22,* 1-36.

Shavelson, R. J., McDonnell, L. M., Oakes, J. (1989). (Eds). *Indicators for Monitoring Mathematics and Science Education.* Santa Monica, CA: The RAND Corporation.

Steen, L. (Ed.). (1987). *Calculus for a New Century* . Washington, DC: Mathematical Association of America.

Stenmark, J. K. (1989). *Assessment Alternatives in Mathematics.* Berkeley: University of California.

Stenmark, J. K. (Ed.). (1991). *Mathematics Assessment: Myths, Models, Good Questions, and Practical Suggestions.* Reston, VA: National Council of Teachers of Mathematics.

Swan, M. (Ed.). (1985). *The Language of Functions and Graphs.* Nottingham, England: Shell Centre for Mathematical Education, University of Nottingham.

Thompson, P. (1994) Students, functions, and the undergraduate curriculum. In E. Dubinsky, A. Schoenfeld, and J. Kaput (Eds.), *Research in Collegiate Mathematics Education,* pp. 21-44. Washington, DC: American Mathematical Society for the Conference Board of the Mathematical Sciences.

Thornton, R. K. (in press). Using large-scale classroom research to study student conceptual learning in mechanics and to develop new approaches to learning. NATO Science Series. Berlin-Heidelberg-New York: Springer Verlag.

Tucker, A. and Leitzel, J. (1995). *Assessing Calculus Reform Efforts.* Washington, DC: Mathematical Association of America.

Tucker, T. (1990). *Priming the Calculus Pump: Innovations and Resources.* Washington, DC: Mathematical Association of America.

Vermont Department of Education. (1992). *Looking Beyond "The Answer."* Montpelier, VT: Author.

Wiggins, G. (1989). A true test: Toward more authentic and equitable assessment. *Phi Delta Kappan, 70* (9), 703-713.

# Appendix

Awards in the
NSF Calculus Program

# PROJECT ABSTRACTS: FY 1988 AWARDS

## An Integrated Program in Calculus and Physics

F. Richard Yeatts      Award No: USE 8813784
Colorado School of Mines      FY 88 $ 74,517
Golden, C0 80401      FY 89 $ 47,451
     FY 90 $ 83,058

An integrated calculus and physics course is being developed and tested. A unique feature is the well-planned laboratory/workshop session where much of the integration of the subject matter occurs. The sessions provide students with the opportunity to explore and discover the relationships between a physical situation, its graphical or geometric representation, and the corresponding analytical representation. The workshop exercises consist of physics experiments, numerical simulations, symbolic manipulations, computer programming, and formal reasoning exercises. Study guides, problem sets, and modular materials are being developed. The students' progress is being evaluated internally as well as externally by a National Advisory Committee.

## The Design of a Computer Algebra System to Effect a More Relevant Mathematics Curriculum

J. Douglas Child      Award No: USE 8814048
Rollins College      FY 88 $ 73,436
Winter Park, FL 32789      FY 89 $ 43,126
     FY 90 $ 44,076

The focus of the three-year project is the construction of a computer environment consisting of a computer algebra system, MAPLE, a specially designed interface to MAPLE, a hypertext system, and other software that is more suitable for teaching and learning calculus for the average student. The computer algebra system demonstrates the reasoning processes of experts. The intent is that students will learn how to think about solving calculus as well as how to solve problems with the help of a computer algebra system. The computer environment is suitable for pre-calculus, science, and engineering curriculum designs. A computer algebra system, both to do and to teach calculus with MAPLE, is being developed along with curriculum and interfaces for computer algebra systems. TM calculus topics are being reordered introducing differentiation and integration early in the course via applied problems. Emphasis is being placed on logic, precise use of language, numerical methods, approximations, and mathematical modeling. Experimental use of materials is taking place at colleges and local high schools which have classes of approximately 35 students. National dissemination is in the form of text to be published by Wadswortb/Brooks-Cole Publishers.

## Calculus Workshops and Conferences

Shair Ahmad      Award No: USE 8813860
University of Miami      FY 88 $ 45,000
Coral Gables, FL 33124      12 months

The project is developing a monthly series of two-day conferences and workshops on calculus to be attended by university and community college faculty members, high school calculus instructors, and industrial representatives. The seminars concentrate on the role of computers and calculators, textbooks, relevance to other disciplines, conceptual understanding, and the development of exercises that stress current technology. Discussions are led by small groups of well-prepared individuals familiar with existing literature on the subjects. Participants are encouraged to carry on similar discussions in their own institutions.

## Calculus Planning Project

Nagambal D. Shah      Award No: USE 8813792
Spelman College      FY 88 $ 50,000
Atlanta, GA 30314      12 months

The project is planning a series of seminars focusing on the special needs that women have in the study of calculus and mathematics, and culminating in a faculty retreat and the writing of a report on the faculty members' shared experiences. The project is being held at a Black institution for females with strong academic traditions and includes formal participation by the chairman of the mathematics department at Agnes Scott College, a female institution with equally strong academic traditions. The program starts with three seminars lead by consultants with special expertise in the area of female studies in mathematics. The presentations and consultations should sensitize the faculty participants to the special elements of females studying mathematics. Three additional consultants will assist with three later seminars that concentrate on the role of computers in the study of calculus. A pilot section of calculus uses and investigates computer software specially designed to assist in the study of calculus. Three undergraduate students help with the evaluation of the materials. With the help of the specialized consultants, the faculty becomes sensitive to the issues involved in the instruction of female students.

### Calculus Curriculum Development

Gerald J. Janusz      Award No: USE 8813873
University of Illinois      FY 88 $ 40,785
Urbana, IL 61801      12 months

Mathematicians are investigating methods by which the teaching of calculus can be made more effective in conveying to students an understanding of calculus as a powerful problem-solving tool. Course material planning focuses on the development of the problem sets that lead students through the central ideas and methods of calculus and enhance their ability to read and write mathematics. The project consults with user departments in science, engineering, and other areas; develops and tests course material for Calculus I based on an approach of Artin; focuses on computational and problem solving; and develops student capacity to read and write mathematics correctly and coherently. A weekly Calculus Seminar trains teaching assistants and discusses content among faculty within and outside the mathematics department. A Calculus Workshop for faculty and high school teachers is being held.

### Calculus, Concepts and Computers

Edward L. Dubinsky      Award No: USE 8813996
Purdue University      FY 88 $ 30,000
West Lafayette, IN 47907      12 months

Three mathematicians are teaching three small (25-30 students) prototype calculus courses based on computer, computer languages, and algebra systems. The geometric and conceptual aspects of calculus, solution of applied problems, and reduction of routine drill by using symbolic manipulation are emphasized. These courses include both mainstream and non-mainstream calculus. The extent of the use of computer labs and the use of the ISTEL and Maple software packages varies. A consulting board of 25 experts from the various academic disciplines will suggest applications from science and mathematics from the latter third of the 20th century. A unique component of the project is research on how students come to understand the underlying ideas in calculus. Theoretical analysis, observations, and experiments on the teaching and learning of calculus are being formulated.

### Planning for a Revitalization of an Engineering/Physical Science Calculus

Elgin H. Johnston      Award No: USE 8813895
Iowa State University      FY 88 $ 49,954
Ames, IA 50011      12 months

Mathematicians are revising the engineering calculus sequence by incorporating modeling and symbolic/ graphical/ numerical software into the curriculum. The planning is done by a committee of faculty from engineering, physical, and mathematical sciences. A calculus network of high school, community college, and college university is being established. Over 90% of the calculus students are from the client disciplines. The project initiates change in a deliberate and timely manner with concurrence by the client departments. One fourth of the first-semester calculus courses are taught under the revised curriculum requiring a small amount of programming, and stressing algorithms and sharply focused real applications. The second and third semesters of calculus are being revised within the same mathematical framework.

### Dynamic Calculus

Robert L. Devaney      Award No: USE 8813865
Boston University      FY 88 $ 40,306
Boston, MA 02215      12 months

An expert fluid dynamicist is developing instructional modules, which incorporate ideas from modem dynamical systems theory into the standard introductory calculus course. The purpose of the project is to augment the calculus with topics of current research interest. Materials can be introduced early in a calculus sequence so that students receive early exposure to topics of contemporary research interest in mathematics, computer experimentation in mathematics, and exciting mathematical visual images. Several modules which show how certain topics in dynamics may be integrated into calculus, and the role of dynamical calculus in science are being developed.

### Calculus in Context

James Callahan      Award No: USE 8814004
Five Colleges, Inc.      FY 88 $ 141,707
Amherst, MA 01002      FY 89 $ 190,845
     FY 90 $ 174,183
     FY 92 $ 74,128

Mathematicians are restructuring the standard three-semester calculus sequence. A new curriculum is being developed in which the four mathematical themes of optimization, estimation and approximation, differential equations, and functions of several variables are stressed from the beginning. These major mathematical concepts grow out of exploring significant problems from social, life, and physical sciences. Dynamical systems, discrete time models, Fourier series, and partial differential equations are some of the concepts which are explored. The computer is being integrated into the curriculum as a basic conceptual device for structuring the way students think about problems and what it means to solve them. Dissemination is in the form of team-taught courses, weekend retreats, summer workshops for area faculty and high school teachers, and publication of the curriculum. These instructional materials are used at universities, liberal arts colleges, and high schools.

### The Language of Change. A Project to Rejuvenate Calculus Instruction

Andrew M. Gleason          Award No: USE 8813997
Harvard University                     FY 88 $ 20,362
Cambridge, MA 02138                    12 months

A group of mathematicians is designing calculus syllabi outlines. They are investigating the use of computers/calculators in opening up new topics and new ways of teaching. They are completely rethinking the goals and content of calculus courses to establish closer collaboration with representatives of client disciplines; to plan the creation of tests; and to plan the development of materials to be used in workshops on pedagogy.

### Calculus Reform in Liberal Arts College

A. Wayne Roberts          Award No: USE 8813914
Macalester College                    FY 88 $ 62,650
Saint Paul, MN 55105                   12 months

Mathematicians are developing a one-year mainstream calculus course. The curriculum stresses basic concepts; numeric and graphic experiments to better understand the power and limitations of technology; the role that calculus plays in changing people's world view; the art of writing a deductive argument; and applied mathematics as a creative modeling process. Outlines of teaching resources to create a lean and lively one-year calculus course include sequences of laboratory-style problems; textbook-type problems for computers/calculators; application modules; open-ended problems; and historical vignettes.

### Calculus. Restructuring and Integration with Computing

Richard H. Crowell          Award No: USE 8814009
Dartmouth College                     FY 88 $ 50,464
Hanover, NH 03755                      12 months

Mathematicians are integrating computers into, and planning the restructuring of, the calculus curriculum. The approach emphasizes elementary functions and the use of the computer for graphical displays and computation of tables of function values. Student-written programs are being used to investigate these functions. Differentiation and integration are taught by means of the difference calculus, making heavy use of the computer. These concepts are used to solve "real-world" problems. Students are expected

to gain a deeper understanding of calculus concepts from the combination of theory, applications, and computer investigations. During the four-course calculus sequence, students are developing the capabilities of doing their own numerical and graphical investigations independently. Text and computer materials, demonstration programs, and problems are being developed. The new curriculum is being tested and evaluated.

### From Euclid to von Neumann, an Activity-Based Learning Experience in Calculus: Project ENABLE

Joan Ferrini-Mundy          Award No: USE 8814057
University of New Hampshire           FY 88 $ 40,487
Durham, NH 03824                       12 months

Mathematicians, mathematics educators, engineers, scientists, high school teachers, and Technical Education Research Center are developing and refining mathematical, educational, and technological perspectives for a three semester calculus curriculum. The project is first conducting a baseline assessment of first-semester calculus students to determine their algebra and trigonometry skills, as well as their understanding of essential precalculus concepts. Implementation of the reorganized and streamlined curriculum requires a clear perception of the students' knowledge base and misconceptions, as well as the students' active participation in their own learning. The curriculum starts with the concept of approximation, whose idealization will lead to derivatives, integrals, and continuity. Biweekly seminars to develop prototype materials, outline modules, and core units are being held during the academic year. Some components are being tested in high schools.

### Student Research Projects in the Calculus Curriculum

Marcus S. Cohen          Award No: USE 8813904
New Mexico State University           FY 88 $ 83,572
Las Cruces, NM 88003                  FY 89 $ 98,993
                                      FY 90 $ 59,652

Mathematicians continue to develop and implement a plan using student research projects in a broad range of calculus courses. Three individual two-week projects are being used instead of hour exams in 35 calculus sections. Almost one-half of the students are minorities. A collection of 400 problems, many annotated with information on their success in the classroom, is being compiled. The projects require that students think broadly and deeply, identifying background material, and synthesizing an approach. Scientists, engineers, and economists help design projects which demonstrate the mathematical underpinnings of solutions to applied problems. Faculty workshops, training of teaching graduate assistants, Advisory Committee meetings, and an extensive evaluation which includes evaluating long-term intellectual growth of students are all taking place as part of the project.

### Planning a Problems-Based Calculus Curriculum

Stephen R. Hilbert                    Award No: USE 8814177
Ithaca College                              FY 88 $ 50,193
Ithaca, NY 14850                              18 months

Mathematicians are developing a problem-based mainstream calculus curriculum. The complex problems will require a minimum of several weeks to solve. The structure of the problems varies from "case study" to "open-ended". Groups of students working together will solve problems which develop essential parts of the calculus, and use calculator/computers where relevant. In-depth interviews with 25 faculty members from accounting, biology, chemistry, economics, finance, management, politics, psychology, and physics help identify realistic problems. Two experimental course sections of Calculus I are being taught, and a one-day conference on the "Future of Calculus" is being held.

### Calculus and the Computer. Innovative Teaching and Learning

William E. Boyce                    Award No: USE 8814011
Rensselaer Polytechnic Institute          FY 88 $ 50,000
Troy, NY 12180-3590                          12 months

Applied mathematicians are developing a calculus course sequence in which computer technology is used to equip students with powerful and versatile problem-solving tools in order to gain deeper understanding of the underlying mathematical concepts. Content includes numerical computation, sophisticated graphics, symbolic computation, relations between mathematics and the natural world, and mathematical modeling. A team of mathematicians, a physicist, an electrical engineer, and students are developing instructional materials to support the use of the computer in calculus. These materials are used in two or three pilot sections of the calculus sequence.

### Development of Calculus

Lawrence C. Moore                    Award No: USE 8814083
Duke University                            FY 88 $ 20,000
Durham, NC 27706                             12 months

A detailed syllabus for a new calculus curriculum is being developed in a cooperative venture with an area high school. The schools are experimenting with the use of computer algebra systems and preparing sample modules. A small

prototype calculus laboratory is being run for the development and testing of interactive experiments and writing as a learning tool in mathematics. Central themes include appropriate use of mathematical and physical tools, identification of a concept and its inverse, use of transformations, and relationships between calculus and real world problems.

### Toward a Conceptual and Captivating Calculus

Thomas A. Farmer                    Award No: USE 8813786
Miami University Oxford Campus             FY 88 $ 48,595
Oxford, OH 45056                             12 months

The project is developing a lean and lively calculus syllabus for college students who have had calculus in high school. After consultation with scientists from client disciplines on the current uses of calculus, materials incorporating computers are being developed to run a preliminary experiment. A large-scale, controlled experiment with teaching and computing materials is being prepared.

### Plan for Calculators in the Calculus Curriculum

Thomas Dick                         Award No: USE 8813785
Oregon State University                   FY 88 $ 27,401
Corvallis, OR 97331-5503                     12 months

A calculus curriculum which makes essential use of the HP-28 symbolic/graphical calculator is being developed and implemented. The objectives are: to identify calculus topics pedagogically suited for use on symbolic/graphical calculators; to identify roles of symbolic/graphical calculators in calculus; and the production of curriculum/calculator materials to be tested during the academic year. A calculus book based on the symbolic/graphical calculator provides both technical advice regarding the calculator and adaptation of the calculus text materials. A series of workshops on utilizing the symbolic/graphical calculator in mathematics classes is being presented.

### Integrated Calculus Development

Alain Schremmer Award                 No: USE 8814000
Community College of Philadelphia          FY 88 $ 40,124
Philadelphia, PA 19107                       12 months

The project is developing a Lagrangian calculus program for the students who are predominantly women, minorities, and returning adults. Lagrangian calculus develops concepts via polynomial approximations rather than limits. It reduces questions about "any" function to the same question about a power function, which appears in the approximating polynomial. The two-semester course is equivalent to pre-calculus and one semester of calculus. The first semester consists of linear approximations, quadratic, and power functions. The second semester consists of the differential study of polynomials, Laurent polynomials, and rational and elementary transcendental functions by Lagrange's approach.

### Revitalization of Calculus

| | |
|---|---|
| Mary McCammon | Award No: USE 8813779 |
| Pennsylvania State University | FY 88 $42,399 |
| University Park, PA 16801 | 12 months |

Mathematics faculty members are developing a lean and lively syllabus for a freshman science and engineering calculus sequence. In consultation with other mathematicians, scientists, and engineers, a core of essential material is being determined. Existing software, computer technology, and placement tests are evaluated and modified, as needed. A test which contains related software and supplements for instructors and which covers the central core of materials is being produced. Several aspects of the experimental syllabus are being taught by the instructors. Each participant experiments with only a small part of the curriculum. In this way, several content areas and approaches can be tested, while insuring that students are exposed to nearly all of the traditional calculus.

### Proposal for a Newsletter on Collegiate Mathematics Education

| | |
|---|---|
| James H. Voytuk | Award No: USE 8814683 |
| American Mathematical Society | FY 88 $104,413 |
| Providence, RI 02901 | FY 89 $66,675 |
| | FY 90 $9,358 |
| | FY 91 $9,358 |

A collegiate mathematics education newsletter is being established. Its purpose is to stimulate greater communication between research mathematicians and collegiate mathematics educators. The newsletter provides a balance of short, timely items directing readers to sources of further information, and longer, more substantive articles presenting discussion of important issues in collegiate mathematics education. The newsletter includes the following: articles on mathematics curriculum; innovative teaching methods funding for collegiate mathematics education; outside classroom activities; profiles of successful mathematics programs; information on conferences, workshops, courses, and use of technology; review of international activities; review of information in other publications; and a column for queries.

### The Calculus Companion: A Computerized Tutor and Computational Aid

| | |
|---|---|
| Edmund A. Lamagna | Award NO: USE 8814017 |
| University of Rhode Island | FY 88 $51,350 |
| Kingston, RI 02881 | 12 months |

The project is creating a computational environment in which calculus students use the computer as both a tutoring device and a computational aid. The system consists of two components: (1) a powerful user interface to a symbolic mathematics package and graphical display routines; and (2) tutorial modules. A study on how the computer can be best integrated into the calculus curriculum, and a prototypical course module on the topic of integration are being completed. Students are introduced to important techniques using symbolic computation facilities. The graphical and numerical capabilities demonstrate several numerical integration techniques. Real world examples from several client disciplines are used to motivate topics.

### Restructuring One Variable Calculus within a Modeling and Computer Oriented Environment

| | |
|---|---|
| Daniel C. Sloughter | Award No: USE 8813781 |
| Furman University | FY 88 $22,476 |
| Greenville, SC 29613 | 12 months |

Mathematicians are developing and testing an experimental one-variable calculus course which builds and analyzes realistic models of dynamic processes, including "chaos". The course restructuring starts with sequences of real numbers and difference equations, and ends with differential equations. Global and qualitative behavior is stressed by use of the computer. There is modeling with symbol manipulation, discrete mathematics, and numerical mathematical packages.

### Development of Computer-Based Curriculum Materials for Calculus: A Planning Project

| | |
|---|---|
| Michael E. Moody | Award No: USE 8814131 |
| Washington State University | FY 88 $29,716 |
| Pullman, WA 99164-3140 | 12 months |

Mathematicians are developing, coordinating, and writing multi-disciplined computer-based curriculum materials for calculus to be implemented at two high schools, a community college, a private college, and a public university. Generic curriculum materials for engineering calculus, calculus for life sciences, and business calculus are being developed by faculty from engineering, biology, chemistry, business, and sociology. The materials include laboratory exercises that use computing devices such as HP-28 and microcomputers using symbolic manipulation programs. These realistic problems use numerical methods that illustrate the power, difficulties, and logic of computation and graphical solution to problems. Electronic "'slide shows" with both animated and static computer graphics of classroom demonstrations and lectures are also being developed.

# PROJECT ABSTRACTS: FY 1989 AWARDS

## Curriculum Development Project.- Calculus

David O. Lomen
University of Arizona
Tucson, AZ 85721

Award No: USE 8953930
FY 89 $ 104,806
12 months

Materials are being developed that complement the calculus courses at major universities throughout the nation. Integrated supplements are being developed that feature laboratories, projects, problems, and software packages. Laboratories are modeled after a typical physics or chemistry laboratory where the student performs guided experiments independent of the present class material. Projects involve the students' discovering and conjecturing results related to calculus. Problems are challenging, realistic questions that might require modem technology to solve. All problems are technology dependent, but independent of a specific brand of computer. The software packages will bridge this gap by supplying the appropriate materials for MS-DOS and Macintosh machines.

## Calculus and Computers: Toward a Curriculum for the 1990s

Marcia C. Linn
University of California
Berkeley, CA 94720

Award No: USE 8953974
FY 89 $ 42,898
12 months

Faculty from a broad spectrum of institutions, including two-year colleges, are learning ways to use Mathematica and exchanging ideas on how to use this powerful tool in the teaching of calculus. The invited speakers at the conference and the PIs are using Mathematica and other integrated symbol manipulation and graphics systems in their calculus courses and are seeing exciting possibilities for their use. The conference participants are learning about these systems and are making suggestions about ways to use these tools.

## Rapid Dissemination of New Calculus Projects

Thomas W. Tucker
Mathematical Association of America
Washington, DC 20001-0000

Award No: USE 8953912
FY 89 $ 41,540
12 months

Detailed descriptions (syllabi, assignments, laboratories, exams, sample text material, preliminary assessment) of eight to 10 new calculus projects are being prepared for publication. Project summaries of approximately 50 additional projects are included.

## Calculus Curriculum Development

J. Jerry Uhl
University of Illinois
Urbana, IL 61801

Award No: USE 8953906
FY 89 $ 25,000
12 months

A non-traditional, entirely new course is being developed under this pilot project through live Mathematica notebooks. Emphasis is placed upon individual student use of the Mathematica program for instruction, computation, and symbolic manipulation within the Mathematica notebooks. The goal is to motivate the students to better understand the foundations and enable them to execute calculations far beyond those expected of students in the traditional course.

## Calculus Redux

Judith H. Morrel
Butler University
Indianapolis, IN 46208

Award No: USE 8953948
FY 89 $ 27,000
12 months

Students are finding more excitement and making better progress in calculus because of a revised curriculum that emphasizes problem-solving, building intuition, and improving written mathematical expression. A data base consisting of non-routine, open-ended, multi-step problems and discussion modules emphasizing concepts, experimentation, and widely varying applications is being created.

## A Revitalization of an Engineering/Physical Science Calculus

Elgin H. Johnston
Iowa State University
Ames, IA 50011

Award No: USE 8953949
FY 89 $ 63,600
FY 90 $ 72,250
FY 91 $ 58,565
FY 92 $ 15,000

A four-year program is under way to revitalize the calculus course taken by science, engineering, and mathematics students. The revised curriculum stresses the modeling and problem-solving aspects of calculus, and teaches students to use commercially available symbolic and numerical software to handle the technical aspects of the subject. Planning, testing, and implementation of the new curriculum are being done under the guidance of a liaison committee made up of faculty from the physical sciences, engineering, and mathematics departments.

### Calculus with Computing A National Model Course

| | |
|---|---|
| Keith D. Stroyan | Award No: USE 8953937 |
| University of Iowa | FY 89 $ 65,000 |
| Iowa City, IA 52242 | 12 months |

The curriculum is being developed to present calculus as the language of science. Beginning calculus is being treated as a laboratory course with modem computers and scientific software as the laboratory equipment. The development is built on a long history of successful use of computers in a calculus laboratory and will make use of new software so that students have a serious start on their education in scientific computation.

### Core Calculus Consortium: A Nationwide Project

| | |
|---|---|
| Andrew M. Gleason | Award No: USE 8953923 |
| Harvard University | FY 89 $ 346,245 |
| Cambridge, MA 02138 | FY 90 $570,283 |
| | FY 91 $ 335,223 |
| | FY 92 $ 418,372 |
| | FY 93 $ 337,500 |

A National consortium of institutions is developing an innovative core calculus curriculum that is practical and attractive to a multitude of institutions. The refocus of calculus uses the "Rule of Three" whereby topics are explored graphically, numerically, and analytically. The consortium is led by Harvard University and consists of the University of Arizona, Colgate University, Haverford-Bryn Mawr Colleges, the University of Southern Mississippi, Stanford University, Suffolk Community College, and Chelmsford High School.

### Calculus Reform in Liberal Arts College

| | |
|---|---|
| A. Wayne Roberts | Award No: USE 8953947 |
| Macalester College | FY 89 $ 199,203 |
| Saint Paul, MN 55105 | FY 90 $ 215,168 |
| | FY 91 $ 148,500 |

A calculus curriculum is being developed that stresses understanding rather than techniques, contains realistic applications, and promotes the ability to write coherent arguments. This development is being carried out with the participation of 26 liberal arts colleges in the Midwest and takes the form of five Resource Collections containing fundamental materials that can be used in part or in total for curriculum development in calculus at any institution. These collections are to be published as five separate volumes.

### The St. Olaf Conference, October 20-22, 1989

| | |
|---|---|
| Paul D. Humke | Award No: USE 8955091 |
| Saint Olaf College | FY 89 $ 1,500 |
| Northfield, MN 55057 | 6 months |

Mathematicians experienced in using computer algebra systems in teaching calculus are meeting to discuss the past experience and their plans for future use. The focus is on how these systems have changed, can change, and will change the teaching of calculus.

### Utilization of Technology in Non-traditional Calculus

| | |
|---|---|
| Wanda Dixon | Award No: USE 8953931 |
| Meridian Community College | FY 89 $ 25,000 |
| Meridian, MS 39301 | 18 months |

The calculus curriculum is being revised to place more emphasis on learning the concepts, solving realistic problems, and improving estimation of skills. Materials are being developed to utilize the HP-285 hand-held calculator.

### Calculus: Restructuring and Integration with Computing

| | |
|---|---|
| Richard H. Crowell | Award No: USE 8953908 |
| Dartmouth College | FY 89 $ 289,171 |
| Hanover, NH 03755 | |

The calculus curriculum is being restructured by integrating into the syllabus student use of a personal computer as a working tool. A substantial body of new courseware is being created that enables the students to use a personal computer as a regular part of their homework to explore, analyze, or verify the central concepts of the calculus, is being created. Students are expected to write some of their own software and/or to modify existing software as an integral part of the course. The course materials are being substantially restructured in order to incorporate the advantages which the presence of the computer affords. The ultimate goal is to produce a new computer-based calculus text.

### C4L Calculus Computers, Calculators and Collaborative Learning

| | |
|---|---|
| Patricia R. Wilkinson | Award No: USE 8953959 |
| CUNY Borough of Manhattan | FY 89 $ 50,000 |
| Community College | 24 months |
| New York, NY 10007 | |

The collaborative learning project is providing students, especially those from minority groups, a better chance to achieve success in calculus. The students are working in informal study groups with the assistance of specially trained tutors.

### *Calculus in the Liberal Arts Curriculum! Multidisciplinary Resources for College Calculus*

| | |
|---|---|
| Ronald W. Jorgensen | Award No: USE 8953926 |
| Nazareth College of Rochester | FY 89 $ 78,232 |
| Rochester, NY 14610 | 24 months |

A calculus curriculum that uses the computer algebra system MAPLE in conjunction with writing assignments that are designed to promote student learning is being developed. The courses are organized in diagnostic learning units and require students to keep a journal which is regularly evaluated by the instructor. This system of ongoing feedback between student and teacher enhances self-evaluation on the part of the student.

### *The Computer Revolution in Calculus: Innovative Approaches to Concepts and Applications*

| | |
|---|---|
| William E. Boyce | Award No: USE 8953904 |
| Rensselaer Polytechnic Institute | FY 89 $ 55,000 |
| Troy, NY 12180-3590 | 12 months |

A new calculus course that exploits the power of a computer as an integral part of teaching and learning is being designed. Advantage is being taken of a computer's capacity to perform numerical computation, produce sophisticated graphics, and carry out extensive symbolic manipulations. Students are provided with powerful and versatile problem-solving tools and simultaneously gain a deeper understanding of the underlying mathematical concepts.

### *Project CALC: Calculus as a Laboratory Course*

| | |
|---|---|
| Lawrence C. Moore | Award No: USE 8953961 |
| Duke University | FY 89 $ 198,522 |
| Durham, NC 27706 | FY 90 $ 217,773 |
| | FY 91 $ 134,570 |
| | FY 92 $ 39,999 |

Students are benefiting from a completely restructured calculus curriculum. The new curriculum features an integrated computer laboratory for exploration and development of intuition, and emphasizes writing to promote student comprehension and expression.

### *Calculators in the Calculus Curriculum*

| | |
|---|---|
| Thomas Dick | Award No: USE 8953938 |
| Oregon State University | FY 89 $ 84,219 |
| Corvallis, OR 97331-5503 | FY 90 $ 75,371 |
| | FY 91 $ 87,918 |
| | FY 92 $ 15,000 |

Calculus students are benefiting from the joint effort involving universities, two- and four-year colleges, high schools, and high technology industry to develop and implement a new calculus curriculum which makes integral use of symbolic/graphical calculators. Text materials appropriate for the equivalent of three semesters of calculus are being produced and classroom-tested in a variety of instructional settings. Workshops provide continuing instructional support for teachers using the curriculum materials and symbolic/graphical calculator.

### *The Calculus Companion: A Computational Environment for Exploring Mathematics*

| | |
|---|---|
| Edmund A. Lamagna | Award No: USE 8953939 |
| University of Rhode Island | FY 89 $ 161,535 |
| Kingston, RI 02881 | 36 months |

The calculus curriculum is being revised to provide students with more complex, real-world problems, to help them develop the skills involved in performing multi-step reasoning, and to help them learn to express mathematical ideas precisely and coherently. A unique computational environment is being developed in which students use the computer as both a tutoring device and a computational aid. The system, called the Calculus Companion, consists of a user-friendly interface to the computer algebra system MAPLE and numerical computation and graphical display routines.

# PROJECT ABSTRACTS: FY 1990 AWARDS

### Software and Project Development for the Two- Year Calculus sequence

David O. Lomen      Award No: USE 9053431
University of Arizona      FY 90 $ 80,000
Tucson, AZ 85721      FY 91 $ 90,000

Calculus students at many institutions are benefiting from integrated supplements: laboratory exercises, projects, problems, and software packages. The laboratory exercises are modeled after a typical chemistry or physics laboratory; projects involve the student discovering and conjecturing results, and problems are challenging and realistic. The software packages provide materials and facilities, and run on MS-DOS or Apple Macintosh machines.

### Implementing Calculus Reform: Conferences, Classroom Testing, and Dissemination

Michael R. Colvin      Award No: USE 9053404
California Polytechnic State University      FY 90 $ 46,996
San Luis Obispo, CA 93407      12 months

A forum is being established for the dissemination and classroom testing of innovative ideas, pedagogy, and technological advances in teaching calculus. Classroom testing is taking place on campuses under the auspices of the California Calculus Consortium. Faculty are introduced to innovative approaches via a summer workshop, and these approaches are reinforced through follow-up activities during the academic year.

### Computer Simulated Experiments in Differential Equations

David A. Horowitz      Award No: USE 9053390
Golden West College      FY 90 $ 36,000
Huntington Beach, CA 92647      12 months

Calculus students are improving their understanding of applied mathematics with the help of computer simulation programs that pictorially and graphically model real-life applications. The package includes growth and decay simulations and harmonic motion simulations.

### Computer Projects and Software for the Introductory Linear Algebra Course

Gareth Williams      Award No: USE 9053365
Stetson University      FY 90 $ 24,964
Deland, FL 32720      12 months

The introductory linear algebra curriculum that is typically part of the two-year calculus sequence is being revitalized by the introduction of computer projects and related material. The projects are diverse in nature, ranging from those that explore mathematical concepts to those that involve mathematical models.

### Calculus and Mathematica

J. Jerry Uhl      Award No: USE 9053372
University of Illinois      FY 90 $ 87,501
Urbana, IL 61801      FY 91 $ 90,004

Students are discovering the concepts and ideas of calculus by exploration and experimentation in a revitalized calculus course, Calculus & Mathematica, that combines the concept of calculus as a laboratory science with correct mathematical foundations. The classroom/laboratory is equipped with Macintosh computers, and the text materials are presented via the Notebook feature of Mathematica.

### Calculus, Concepts, and Computers

Edward L. Dubinsky      Award No: USE 9053432
Purdue University      FY 90 $ 220,000
West Lafayette, IN 47907      FY 91 $ 226,000
     FY 92 $ 200,000

Students are learning both the geometric aspects of calculus using computer graphics and the mathematical concepts via a mathematical programming language that allows them to make standard mathematical constructions using standard mathematical notation; drill and practice are being reduced by using a computer algebra system. Research into the process of learning the underlying ideas of calculus is also being conducted.

### Engineering/Physical Science Second Year Calculus and Differential Equations. A Pilot Project

Leslie Hogben     Award No: USE 9053428
Iowa State University     FY 90 $ 48,241
Ames, IA 50011     12 months

Students in fourth-semester calculus, differential equations, and linear algebra are benefiting from a revised curriculum that presents the underlying mathematics by introducing physical problems which require mathematics for their solution.

### Calculus with Computing: A National Model Course

Keith D. Stroyan     Award No: USE 9053383
University of Iowa     FY 90 $ 64,000
Iowa City, IA 52242     FY 91 $ 66,000

Students in the Accelerated Calculus Program are benefiting from a new curriculum that treats beginning calculus as a laboratory course with NeXT computers and Mathematica software as the equipment. Students, in classes of 125, work on open-ended projects with assistance from graduate teaching assistants, upper-class undergraduates, and faculty.

### A Reformed Calculus Program Based on Mathematics and Project CALC

William H. Barker     Award No: USE 9053397
Bowdoin College     FY 90 $ 35,000
Brunswick, ME 04011     FY 91 $ 44,000

Students are learning calculus in a discovery-based laboratory course using materials developed at Duke University and adapted for use in a liberal arts college setting. The course and laboratory materials are made available for Macintosh computers and exploit the Notebook feature of the computer algebra system, Mathematica.

### Computer Algebra System Workshops, New Series

Donald B. Small     Award No: USE 9053427
Colby College     FY 90 $ 84,875
Waterville, ME 04901     FY 91 $ 10,000

College and pre-college faculty are learning how to use computer algebra systems in the teaching of calculus. Strong emphasis is placed on using these systems in such a way that the calculus curriculum is improved by introducing numerical and graphical experimentation, and by focusing on problem-solving and understanding of concepts.

### A Workshop on the Undergraduate Linear Algebra Curriculum

David Lay     Award No: USE 9053422
University of Maryland     FY 90 $ 43,922
College Park, MD 20742     12 months

Mathematical Scientists are assessing the current state of linear algebra instruction in the undergraduate curriculum, laying a foundation for its improvement, and identifying priorities for further and continuing study. This endeavor is being carried out by means of a survey of linear algebra curricula, a workshop, and a conference.

### Video Applications Modules in Calculus

Frank R. Giordano     Award No: USE 9053407
Consortium for Mathematics &     FY 90 $ 101,851
Its Applications, Inc.     12 months
Arlington, MA 02174

A calculus video applications library is being produced that will expose students to exciting applications of mathematics, and includes a printed teacher's guide. The modules can also be used in faculty development activities and contain print packages with special attention to showing how to introduce these ideas into the classroom.

### A Modular Calculus

William W. Farr     Award No: USE 9053430
Worcester Polytechnic Institute     FY 90 $ 56,981
Worcester, MA 01609     FY 91 $ 61,572

Students are benefiting from a new calculus curriculum that features early development of multivariable functions and derivatives, a less sequential approach to the calculus topics, and the development of team projects with computer laboratories and written laboratory reports.

### Computers in Calculus, The Dearborn Project

David A. James     Award No: USE 9053385
University of Michigan     FY 90 $ 57,500
Dearborn, MI 48128     FY 91 $ 59,000

Calculus students are benefiting from a new curriculum constructed from the best curriculum development efforts around the country. After classroom testing, the results are being evaluated and a package of computer laboratory materials and an instructor's manual is being desktop-published and disseminated.

*First-Year Calculus From Graphical, Numerical, and Symbolic Points of View*

Arnold M. Ostebee                  Award No: USE 9053363
Saint Olaf College                              FY 90 $ 49,977
Northfield, MN 55057                          FY 91 $ 54,965

A new curriculum combining graphical, numerical, and algebraic viewpoints on the main ideas and objects of calculus and supported by modern computing technology is helping students understand calculus ideas more deeply and apply them more effectively.

*A Model Program Using Student Research Projects in Calculus and Differential Equations*

David J. Pengelley                  Award No: USE 9053387
New Mexico State University              FY 90 $ 120,000
Las Cruces, NM 88003                            24 months

Calculus students are benefiting from a newly designed course in vector calculus and differential equations that equips them with the basic tools of modern mathematical modeling. As a result of analyzing a sequence of discovery projects, students see how real-world questions may engender theoretical tools and how these tools may then be extended to new applications. The independent experimentation, conjecture, and testing this fosters builds the confidence needed for independent work or group leadership. Several national workshops are being held on departmental implementation of a projects-based calculus curriculum, as well as a conference on using discovery projects to teach basic ideas in calculus and differential equations.

*Computer Enhancement Options for Second Year Calculus*

George R. Livesay                  Award No: USE 9053426
Cornell University-Endowed                  FY 90 $ 93,000
Ithaca, NY 14853                              FY 91 $ 97,000

Specialized software packages are being developed for the differential equations and vector and multivariable calculus topics typically included in a second-year calculus course. The new software is modeled after the MacMath and Analyzer packages already completed and classroom tested.

*Developing a Projects-Based Calculus Curriculum*

Stephen R. Hilbert                  Award No: USE 9053416
Ithaca college                              FY 90 $ 86,175
Ithaca, NY 14850                                24 months

Students are achieving increased understanding of concepts, seeing the unity of the important topics in calculus, obtaining a deeper geometric understanding, and learning problem-solving skills in a new calculus course that integrates large, open-ended problems into the curriculum.

*A Laboratory Approach to Calculus*

L. Carl Leinbach                    Award No: USE 9053401
Gettysburg College                          FY 90 $ 59,225
Gettysburg, PA 17325                             8 months

College faculty are learning how computer algebra systems can be used to improve calculus curricula and are designing new curricula that integrates laboratories into the new courses. The new designs being implemented are critiqued and redesigned at a follow-up workshop.

*Duke University's Project CALC Test Site*

Alvin J. Kay                        Award No. USE 9053364
Texas A&I University                        FY 90 $ 17,664
Kingsville, TX 78363                            12 months

Students are learning calculus in a laboratory setting and are discovering for themselves the concepts and problem solving power of calculus. The overall framework of the laboratory and the course materials being used were developed at Duke University and are being tested in this setting.

# PROJECT ABSTRACTS: FY 1991 AWARDS

*Calculus with Computers for the Mid-Sized University: Adapting and Testing the Iowa Materials*

Steven C. Leth      Award No: USE 9153277
University of Northern Colorado      FY 91 $ 45,000
Greeley, CO 80639      FY 92 $ 20,000

The Iowa materials and approach to teaching calculus are being adapted, refined, and implemented throughout the calculus sequence. A lecture approach is being integrated with an interactive computer laboratory component centered around Mathematica Notebooks. Many of the students are future mathematics teachers.

*Integration of Computing into Main-Track Calculus*

James F. Hurley      Award No. USE 9153270
University of Connecticut      FY 91 $ 41,723
Storrs, CT 06268      FY 92 $ 79,907

The three-semester calculus sequence is being revised to integrate the computer as an active component of the learning process. A pilot program is being expanded throughout the three-semester sequence. A laboratory component is being introduced in which students will modify computer code written in True BASIC and apply the programs to a wide range of mathematical problems.

*Connecticut Calculus Consortium*

Robert J. Decker      Award No: USE 9153298
University of Hartford      FY 91 $ 100,000
West Hartford, CT 06117      FY 92 $ 70,000

Students are introduced to realistic problems and to the technology (graphing calculators and microcomputers) that is capable of dealing with them. Existing materials are adapted and implemented on a state-wide basis. The laboratory materials and the text materials are integrated into the new course.

*The Georgia Tech-Clemson Consortium for Undergraduate Mathematics in Science and Engineering*

Alfred D. Andrew      Award No: USE 9153309
Georgia Tech Research Corporation      FY 91 $ 83,560
Georgia Institute of Technology      FY 92 $ 85,991
Atlanta, GA 30332      FY 93 $ 24,417

A large-scale adaptation, refinement, and implementation project is invigorating the teaching and learning of calculus for science and engineering students. Innovations being adapted have been tested at the participating institutions. The three-year effort is taking advantage of both modern supercalculator and microcomputer technology and is incorporating group learning through team projects. The project will affect some 20,000 students over five years.

*Multivariable Calculus Using Mathematica*

Dennis M. Schneider      Award No: USE 9153249
Knox College      FY 91 $ 45,431
Galesburg, IL 61401      FY 92 $ 42,569

Computer technology is being exploited to produce a "leaner and livelier" multi-variable calculus curriculum supported by a collection of Mathematica Notebooks and graphics packages that provide material for classroom use as well as problems that invite students to explore further the concepts of calculus. A text is being produced which assumes that students have access to a powerful computing environment, but not necessarily Mathematica.

*Disseminating Calculus in Context*

James Callahan      Award No: USE 9153301
Five Colleges, Inc.      FY 91 $ 85,000
Amherst, MA 01002      FY 92 $ 120,000

The project is adapting and disseminating the Calculus in Context curriculum. The group of mathematicians with experience with this curriculum is being substantially expanded through intensive workshops, and materials are being prepared that allow instructors in high schools, two-year colleges, and four-year colleges and universities to teach the course without special training. Evaluation of the efficacy of the approach and of the dissemination efforts are part of the project.

### Calculus in a Real and Complex World, Year II

Franklin A. Wattenberg          Award No: USE 9153266
University of Massachusetts                  FY 91 $ 34,205
Amherst, MA 01003                            FY 92 $ 29,562

A new, two-semester sequence is being developed which includes differential equations and linear algebra, and is based on the philosophy and spirit of the first-year calculus course developed under the Five Colleges Calculus in Context Project. Students gain a better grasp of the concepts of calculus when they are presented in the context of real and substantial applications that require a combination of techniques involving open-ended problems that often do not have clean, simple solutions. Writing and the use of computers are an integral part of this approach.

### Calculus Reform at a Comprehensive State University with Project CALC

Charles C. Alexander            Award No: USE 9153283
University of Mississippi                    FY 91 $ 60,000
University, MS 38677                          12 months

The materials and approach to teaching calculus developed at Duke University are being adapted and implemented. The revised course emphasizes greater conceptual understanding through extensive writing, collaborative learning in a discovery-based microcomputer laboratory, and using mathematics for modeling real world phenomena.

### Project to Adapt and Refine Purdue Model for Teaching Calculus for Liberal Arts and State Colleges

Carol L. Freeman                Award No: USE 9153259
Nebraska Wesleyan University                 FY 91 $ 70,000
Lincoln, NE 68504                            24 months

A consortium of institutions are adapting, refining, and implementing the approach to teaching calculus. Issues of computer anxiety and cooperative learning are being examined during the implementation phase. The adaptation phase includes introducing the use of graphing calculators.

### Calculator Enhanced Instruction Project by a Consortium of NJ Community Colleges

Jean Lane                       Award No: USE 9153258
Union County College                         FY 91 $ 77,415
Cranford, NJ 07016                           12 months

Five community colleges are adapting a calculator-based curriculum for calculus. Workshops are being conducted to introduce the faculty to the new curriculum and to begin its implementation. The faculty introduces the new approach to their colleagues as it is implemented in all calculus sections.

### A Problem-Based Restructuring of Calculus

Jacob Barshay                   Award No: USE 9153248
CUNY City College                            FY 91 $ 55,000
New York, NY 10031                           12 months

Students are working collaboratively in small teams under the guidance of advanced undergraduates on thought-provoking problems. The collection of problems is being expanded and includes materials for use with graphing calculators and guide books for the restructured course.

### The Rensselaer-Albany Regional Calculus Consortium- A Curriculum Adaptation, Refinement, and Implementation Program

Timothy L. Lance                Award No: USE 9153252
SUNY at Albany                               FY 91 $ 41,000
Albany, NY 12201                             12 months

A revised calculus curriculum is being adapted, refined, and implemented. The faculty are attending workshops and computer-calculus classes.

### Dissemination of Project CALC Methods and Materials

Lawrence C. Moore               Award No: USE 9153272
Duke University                              FY 91 $ 162,165
Durham, NC 27706                             FY 92 $ 97,294

Third-semester calculus materials are being completed for teaching calculus as a laboratory course. The work includes the expansion of the repertoire of classroom and laboratory projects, development of versions of alternate software and hardware environments, and completion of a high school version of the course. The methods and materials for all three semesters are being evaluated and widely disseminated by workshops for college-level and pre-college faculty, by continued publication of a newsletter, and by production of preliminary materials.

### Calculus & Mathematica at Ohio State

| | |
|---|---|
| William J. Davis | Award No: USE 9153246 |
| Ohio State University | FY 91 $ 99,916 |
| Columbus, OH 43210 | FY 92 $ 91,277 |

The Calculus & Mathematica course is being extended to second-year calculus. The focus is on developing of materials, testing them in the classroom, and revising them in light of the experience gained. Student outcomes are being assessed and the results reported to the community.

### Complete Implementation of a Mathematica Laboratory for Calculus at a Public Metropolitan University

| | |
|---|---|
| Richard E. Mercer | Award No: USE 9153300 |
| Wright State University | FY 91 $ 91,629 |
| Dayton, OH 45435 | 24 months |

Materials are being adapted and new materials developed to implement a laboratory calculus course. The materials emphasize extensive use of Mathematica programming, especially graphics routines and conceptual questions that require written responses in paragraph form. The curricular changes emphasize a systematic treatment of the approximation of functions throughout the calculus sequence.

### Implementation and Adaptation of St. Olaf First-Year Calculus In the Schools of the Chattanooga Consortium

| | |
|---|---|
| Stephen W. Kuhn | Award No: USE 9153285 |
| University of Tennessee | FY 91 $ 65,000 |
| Chattanooga, TN 37403-2598 | FY 92 $ 70,000 |

Faculty at the university, college, and high school levels are working together to improve their curriculum by adapting and implementing the model calculus program being developed. Graphical, numerical, and algebraic viewpoints are brought to bear on calculus ideas to improve students' conceptual understanding.

### Gems of Exposition in Elementary Linear Algebra

| | |
|---|---|
| Charles R. Johnson | Award No: USE 9153284 |
| College of William and Mary | FY 91 $ 92,000 |
| Williamsburg, VA 23185 | 24 months |

Efforts are underway to collect and broadly disseminate gems of exposition in elementary linear algebra. These include especially insightful proofs, short and open-ended problems, longer expositional items, and machine-oriented, computational exercises, all of which are designed to communicate fundamental linear algebra ideas to beginning students. Items are being solicited from a wide range of individuals worldwide and are being published in a volume designed to be available at low cost.

### Multi-HBCU Calculus Project

| | |
|---|---|
| Walter Elias | Award No: USE 9153264 |
| Virginia State University | FY 91 $ 50,000 |
| Petersburg, VA 23803 | FY 92 $ 100,000 |

The calculus curriculum is being revised. A problem-solving approach that encourages experimentation and enhances the study of calculus by minority students is being implemented. The primary technical tool is the microcomputer running the computer algebra system Derive.

### The Washington Center Calculus Dissemination Project

| | |
|---|---|
| Robert S. Cole | Award No: USE 9153274 |
| Evergreen State College | FY 91 $ 80,062 |
| Olympia, WA 98505 | FY 92 $ 145,348 |

Faculty from a consortium of diverse institutions are adapting, refining, and implementing approaches to teaching calculus. A select group of about 20 faculty are being trained in the new methods. This core of faculty trains faculty from twelve additional institutions, and all those involved adapt and implement one of the new approaches at their home institutions.

# PROJECT ABSTRACTS: FY 1992 AWARDS

## Project CALC- Calculus as a Laboratory Course

Lawrence C. Moore
Duke University
Durham, NC 27708

Award No: DUE 9241916
FY 1989 $ 198,522
FY 1990 $ 217,773
FY 1991 $ 134,570
FY 1992 $ 39,999
Calculus

Students are benefiting from a completely restructured calculus curriculum at Duke University and the North Carolina School of Science and Mathematics. The new curriculum features an integrated computer laboratory for exploration and development of intuition and emphasizes writing to promote student comprehension and expression. The course materials are being developed jointly by members of the faculties of the two schools.

## Calculus, Concepts, and Computers

Edward L. Dubinsky
Purdue Univ. Research Foundation
West Lafayette, IN 47907

Award No: DUE 9242137
FY 1990 $ 220,000
FY 1991 $ 226,000
FY 1992 $ 200,000
Calculus

Students are learning the geometric aspects of calculus using computer graphics and are learning the mathematical concepts via a mathematical programming language that allows them to make standard mathematical constructions using standard mathematical notation; drill and practice are being reduced by using a computer algebra system. Research into the process of learning the underlying ideas of calculus is also being conducted.

## Dissemination of Project CALC Methods and Materials

Lawrence C. Moore
Duke University
Durham, NC 27708

Award No: DUE 9244194
FY 1991 $ 97,294
FY 1992 $ 97,294
Calculus

Third-semester calculus materials are being completed for teaching calculus as a laboratory course. The work includes the expansion of the repertoire of classroom and laboratory projects, development of versions of alternate software and hardware environments, and completion of a high school version of the course. The methods and materials for all semesters are being evaluated and widely disseminated by workshops for college-level and precollege faculty, by

continued publication of a newsletter, and by production of preliminary materials.

## Connecticut Calculus Consortium

Robert J. Decker
University of Hartford
West Hartford, CT 0611

Award No: DUE 9244308
FY 1991 $ 100,000
FY 1992 $ 70,000
Calculus

A group of 18 institutions is working together to introduce students to realistic problems and to the technology (graphing calculators and microcomputers) that is capable of dealing with them. Existing materials are being adapted and will be implemented on a state-wide basis. The laboratory materials developed at the University of Hartford and the text materials being developed by the Core Calculus Consortium (led by Harvard University) are being integrated into the new course.

## Calculus in a Real and Complex World Year II

Franklin Wattenberg
University of Massachusetts - Amherst
Amherst, MA 01003

Award No: DUE 9244356
FY 1991 $ 34,205
FY 1992 $ 29,562
Calculus

A new two-semester sequence is being developed that includes differential equations and linear algebra and that is based on the philosophy and spirit of the first-year calculus course developed under the Five Colleges Calculus in Context Project. Students gain a better grasp of the concepts of calculus when they are presented in the context of real and substantial applications that require a combination of techniques and that involve open-ended problems that often do not have clean, simple solutions. Writing and the use of computers are an integral part of this approach.

## Calculus with Computers for the Mid-Sized University: Adapting and Testing the Iowa Materials

Steven C. Leth
University of Northern Colorado
Greeley, CO 80639

Award No: DUE 9244362
FY 1991 $ 45,000
FY 1992 $ 20,000
Calculus

The materials and the approach to teaching calculus developed at the University of Iowa are being adapted, refined, and implemented throughout the calculus sequence. A lecture approach is integrated with an interactive computer laboratory component centered around Mathematica Notebooks. Many of the students are future mathematics teachers.

### The Washington Center Calculus Dissemination Project

| | |
|---|---|
| Robert S. Cole | Award No: DUE 9244364 |
| Evergreen State College | FY 1991 $ 80,062 |
| Olympia, WA 98505 | FY 1992 $ 145,348 |
| | Calculus |

Faculty from a consortium of diverse institutions are adapting, refining, and implementing approaches to teaching calculus that were developed at Duke University and by the Harvard Calculus Consortium. The first year involved training a select group of about 20 faculty in the new methods. This core of faculty trains faculty from 12 additional institutions in the second year, and all those involved adapt and implement one of the new approaches at their home institution.

### Multivariable Calculus Using Mathematica

| | |
|---|---|
| Dennis M. Schneider | Award No: DUE 9244434 |
| Knox College | FY 1991 $ 45,431 |
| Galesburg, Ill 61401 | FY 1992 $ 42,569 |
| | Calculus |

Computer technology is being exploited to produce a "leaner and livelier" multi-variable calculus curriculum supported by a collection of Mathematica Notebooks and graphics packages that provide material for classroom use as well as problems that invite students to explore further the concepts of calculus. A text will result that will assume that students have access to a powerful computing environment, but not necessarily Mathematica.

### Calculus in Context

| | |
|---|---|
| James Callahan | Award No: DUE 9240180 |
| Five Colleges Incorporated | FY 1988 $ 141,707 |
| Amherst, MA 01002 | FY 1989 $ 190,845 |
| | FY 1990 $ 174,183 |
| | FY 1991 $ 129,392 |
| | FY 1992 $ 74,128 |
| | Calculus |

Mathematicians from the Five College Consortium of Amherst, Hampshire, Mount Holyoke, Smith Colleges, and the University of Massachusetts are restructuring the standard three-semester calculus sequence. They are developing a new curriculum in which the four mathematical themes of optimization, estimation and approximation, differential equations, and functions of several variables are stressed. These major mathematical concepts grow out of exploring significant problems from social, life, and physical sciences. Dissemination is in the form of team-taught courses, weekend retreats, summer workshops for area faculty and high school teachers, and publication of the curriculum. These instructional materials will be used at universities, liberal arts colleges, and high schools.

### A Revitalization of an Engineering/Physical Science Calculus

| | |
|---|---|
| Elgin H. Johnston | Award No: DUE 9245079 |
| Iowa State University | FY 1989 $63,600 |
| Ames, IA 50011 | FY 1990 $ 72,250 |
| | FY 1991 $ 58,565 |
| | FY 1992 $ 15,000 |
| | Calculus |

A four-year program is under way to revitalize the calculus course taken by science, engineering, and mathematics students. The revised curriculum stresses the modeling and problem-solving aspects of calculus and teaches students to use commercially available symbolic and numerical software to handle the technical aspects of the subject. The planning, testing, and implementation of the new curriculum are being done under the guidance of a liaison committee made up of faculty from the physical sciences, engineering, and mathematics.

### Disseminating Calculus in Context

| | |
|---|---|
| James Callahan | Award No: DUE 9245065 |
| Five Colleges Incorporated | FY 1991 $ 85,000 |
| Amherst, MA 01002 | FY 1992 $ 120,000 |
| | Calculus |

A two-year project is under way to adapt and disseminate the Calculus in Context curriculum being taught at Hampshire, Mount Holyoke, and Smith Colleges. The group of mathematicians experienced with this curriculum is being substantially expanded through intensive workshops. Materials are being prepared that will allow instructors in high schools, two-year colleges, and four-year colleges and universities to teach the course without special training. Evaluation of the efficacy of the approach and of the dissemination efforts is part of the project.

*Implementation and Adaptation of St. Olaf First- Year Calculus in the Schools of the Chattanooga Consortium*

| | |
|---|---|
| Stephen W. Kuhn | Award No: DUE 9245470 |
| University of Tennessee–Chattanooga | FY 1991 $ 65,000 |
| Chattanooga, IN 37403-2598 | FY 1992 $ 70,000 |
| | Calculus |

Faculty at the University of Tennessee, Chattanooga, the Chattanooga State Technical Community College, Southern College, and area high schools are working together to improve their curriculum by adapting and implementing the model calculus program being developed at St. Olaf College. Graphical, numerical, and algebraic viewpoints are brought to bear on calculus ideas to improve students' conceptual understanding.

*Calculators in the Calculus Curriculum*

| | |
|---|---|
| Thomas P. Dick | Award No: DUE 9245080 |
| Oregon State University | FY 1989 $ 84,219 |
| Corvallis, OR 97331-5503 | FY 1990 $ 75,371 |
| | FY 1991 $ 87,918 |
| | FY 1992 $ 15,000 |
| | Calculus |

Calculus students are benefiting from this joint effort involving universities, two- and four-year colleges, high schools, and high technology industry to develop and implement a new calculus curriculum which makes integral use of symbolic/graphical calculators. Text materials appropriate for the equivalent of three semesters of calculus are being produced; these materials are being class-tested in a variety of instructional settings; workshops are providing instructional support for teachers using the curriculum materials and symbolic/graphical calculator.

*Core Calculus Consortium: A Nationwide Project*

| | |
|---|---|
| Andrew M. Gleason | Award No: DUE 9245088 |
| Harvard University | FY 1989 $ 346,245 |
| Cambridge, MA 02138 | FY 1990 $ 570,283 |
| | FY 1991 $ 335,223 |
| | FY 1992 $ 418,372 |
| | FY 1993 $ 337,500 |
| | Calculus |

A national consortium of institutions is developing an innovative core calculus curriculum that is practical and attractive to a multitude of institutions. The consortium is led by Harvard University and also includes the University of Arizona, Colgate University, Haverford-Bryn Mawr Colleges, the University of Southern Mississippi, Stanford University, Suffolk Community College, and Chelmsford

High School. The refocus of calculus is using the "Rule of Three" whereby topics are explored graphically, numerically, and analytically.

*Calculus and Mathematica at Ohio State*

| | |
|---|---|
| William J. Davis | Award No: DUE 9245469 |
| Ohio State University | FY 1991 $ 99,916 |
| Columbus, OH 43210 | FY 1992 $ 91,277 |
| | Calculus |

The Calculus and Mathematica course initiated at the University of Illinois and further developed at Ohio State is being extended to second-year calculus. The focus is on development of materials, testing them in the classroom, and revising them in light of the experience gained. Student outcomes are being assessed, and results reported to the community.

*Multi-HBCU (Historically Black Colleges and Universities) Calculus Project*

| | |
|---|---|
| Walter Elias | Award No: DUE 9245501 |
| Virginia State University | FY 1991 $ 50,000 |
| Petersburg, VA 23803 | FY 1992 $ 100,000 |
| | Calculus |

Four HBCU's are working together to revise their calculus curriculum. A problem-solving approach is being implemented that encourages experimentation and enhances the study of calculus by minority students. The primary technical tool is the microcomputer running the computer algebra system Derive.

*The Georgia Tech-Clemson Consortium for Undergraduate Mathematics in Science and Engineering*

| | |
|---|---|
| Alfred D. Andrew | Award No: DUE 9245528 |
| Georgia Institute of Technology | FY 1991 $ 83,560 |
| Atlanta, GA 30332 | FY 1992 $ 85,991 |
| | FY 1993 $ 24,417 |
| | Calculus |

A large-scale adaptation, refinement, and implementation project is invigorating teaching and learning calculus for science and engineering students. Innovations being adapted have been tested at each of the two participating institutions and at Iowa State University, New Mexico State University, and Cornell University. The three-year effort is taking advantage of both modern supercalculator and Microcomputer technology and incorporates group learning through team projects. The project will affect some 20,000 students over five years.

### Integration of Computing into Main-Track Calculus

| | |
|---|---|
| James F. Hurley | Award No: DUE 9245548 |
| University of Connecticut | FY 1991 $ 41,723 |
| Storrs, CT 06268 | FY 1992 $ 79,771 |
| | Calculus |

The three-semester calculus sequence is being revised to integrate the computer as an active component of the learning process. A pilot program begun in 1989 is being expanded throughout the three-semester sequence. A laboratory component is being introduced in which students will modify computer code written in True BASIC and apply the programs to a wide range of mathematical problems.

### The Rensselaer-Albany Regional Calculus Consortium-A Curriculum Adaptation, Refinement and Implementation

| | |
|---|---|
| Timothy Lance | Award No: DUE 9246707 |
| SUNY - Albany | FY 1991 $ 41,000 |
| Albany, NY 12201 | FY 1992 $ 16,333 |
| | Calculus |

The University at Albany, in collaboration with Rensselaer Polytechnic Institute, is incorporating interactive adaptation and refinement of two computer-intensive approaches to teaching calculus; the Albany and Rensselaer models from previous years; ongoing training workshops about this learning environment for our own mathematics faculties and those of area two- and four-year colleges and secondary schools; creation of a "virtual classroom" for broader dissemination of our ideas and testing of a model for distance learning of mathematics in a computer-intensive environment; site testing of existing NSF-sponsored computer calculus initiatives. A goal is to create a distributed version of our own local computer classrooms.

### Calculus and Mathematica

| | |
|---|---|
| Jerry Uhl | Award No: DUE 9252484 |
| University of Illinois - Urbana | FY 1992 $ 150,000 |
| Urbana, IL 61801 | FY 1993 $ 150,000 |
| | FY 1994 $ 150,000 |
| | Calculus |

Calculus and Mathematica is a laboratory course in calculus based on electronic interactive notebooks written within the Mathematica system. In two years, the teaching of Calculus and Mathematica has spread to more than 20 colleges and 6 high schools. The proposed work is continuing this project by extending the development, dissemination, and evaluation of the existing Calculus and

Mathematica project together with a pilot development of a Differential Equations course.

### Fully Renewed Calculus at Three Large Universities

| | |
|---|---|
| Keith D. Stroyan | Award No: DUE 9252486 |
| University of Iowa | FY 1992 $ 80,471 |
| Iowa City, LA 52242 | FY 1993 $ 39,751 |
| | FY 1994 $ 42,184 |
| | Calculus |

This collaborative project is being implemented at the University of Iowa, University of Wisconsin - La Crosse, and Brigham Young University. Renewed calculus materials developed at the University of Iowa are being revised, tested, and refined at these three institutions. In addition to learning traditional calculus skills, students are exposed to new ideas through large open-ended projects on a variety of scientific and mathematical projects.

### The Rhode Island Calculus Consortium Module Project

| | |
|---|---|
| Lewis Pakula | Award No: DUE 9252468 |
| University of Rhode Island | FY 1992 $ 150,966 |
| Kingston, RI 02881 | Calculus |

The Rhode Island Calculus Consortium is composed of Rhode Island university, college, and high school faculty who are seeking to introduce new approaches to calculus instruction. It is adapting ideas and materials from successful pilot calculus projects to create a series of self-contained instructional modules for use in the first two semesters of college or high school (AP) calculus. Each module is organized around a set of problems, projects, and examination items, as well as text material and class and group activities. Modules will be piloted, revised, and evaluated by consortium instructors. A fundamental feature of this project is that the modules will be assembled by the consortium members themselves. A Summer Institute and a Calculus Colloquium will disseminate to a wider regional audience the products of the project effort, and with them. the broad aims and ideas of calculus reform.

### The University of Connecticut Computer-integrated Calculus Project

| | |
|---|---|
| James F. Hurley | Award No: DUE 9252463 |
| University of Connecticut | FY 1992 $ 177,746 |
| Storrs, CT 06268 | FY 1993 $ 197,064 |
| | FY 1994 $ 219,434 |
| | Calculus |

Connecticut's program seeks to (1) involve the computer as a tool for fostering in-depth conceptual understanding of both the ideas and techniques of calculus; (2) promote analytical thinking in dealing with quantitative relationships; (3) sharpen students' geometric, numerical, and theoretical intuition, and increase their ability to visualize associated basic relationships; (4) convey the usefulness of and provide practice with the computer as a tool for carrying out mathematical computations; (5) develop in students the habit of working cooperatively with peers to analyze and solve problems, and the confidence to persist in mathematical analysis of complex, multi-step phenomena; and (6) make calculus reform feasible at institutions where cost and ease of transition from a traditional calculus program are significant issues. Dissemination to 78 participating high schools in the University's Cooperative Education Program is under way

### Metrolina Calculus Consortium: Implementing a Technology-based Calculus Curriculum

Mary K. Prichard      Award No: DUE 9252502
University of North Carolina–Charlotte   FY 1992 $153,484
Charlotte, NC 28223      Calculus

The Metrolina Calculus Consortium is designed to facilitate the implementation of different calculus reform materials into schools in the Charlotte, North Carolina, area thorough workshops and related activities. The project includes four components: adaptation of curriculum materials to fit local needs, faculty development, research on student learning, and dissemination. The project provides support and resources for teachers at these institutions to adapt and implement a technology-based calculus curriculum. Research on student learning is a critical component of this project.

### Implementation of Calculus Reform at a Comprehensive State University with Project CALC

C.C. Alexander      Award No: DUE 9252516
University of Mississippi      FY 1992 $ 145,941
University, MS 38677      Calculus

The University of Mississippi is implementing a program of calculus reform using materials and methods of Project CALC. By the conclusion of the proposed phase-in period, all regular faculty in the department will have an opportunity to teach a section of Project CALC. The university will serve as a calculus test site for the Mississippi Alliance for Minority Participation as the alliance strives to transform "gatekeeping" courses into "gateway" courses. One project goal will be to refine the Project CALC materials, and possibly the implementation strategy, based on our experiences (this will include the adaptation of the materials to MathCAD 3.1).

### Preparing for a New Calculus

Anthony L. Peressini      Award No: DUE 9252475
University of Illinois - Urbana      FY 1992 $ 91,394
Urbana, IL 61801      Calculus

The conference/workshop "Preparing for a New Calculus" is bringing together 80 leaders in mathematics curriculum reform, including calculus reform, school mathematics reform in light of the NCTM Curriculum Standards, and educational technology initiatives. The deliberations of this conference will yield a set of action-oriented recommendations focusing on the implications of the emerging calculus courses on school mathematics training, and on how recent developments in content and methods in high school mathematics will impact the calculus courses.

### Calculus Reform in Western Appalachia A Consortium Approach

James H. Wells      Award No: DUE 9252494
University of Kentucky      FY 1992 $ 378,459
Lexington, KY 405060057      FY 1993 $ 324,081
     FY 1994 $ 47,460
     Calculus

Mathematicians at five public and four private institutions in three western Appalachian states (Kentucky, Tennessee, and Virginia) are developing and integrating modem calculus curricula into their instructional programs. In addition to systematically involving departmental colleagues in the teaching of the new curricula, they are developing and implementing training programs for teaching assistants and workshops to familiarize faculty in other disciplines with the philosophy and implications of the revisions. The organized system of collaboration and information-sharing among the participants is integrated into an extensive program of dissemination, including television, directed at colleagues in sister institutions and high schools. They are placing particular emphasis on efforts to inform and involve high school teachers of calculus and precalculus.

### Mid-Atlantic Regional Calculus Consortium

Joshua A. Leslie      Award No: DUE 9252508
Howard University      FY 1992 $ 75,000
Washington, DC 20059      Calculus

MARCC—a consortium of five Black universities, a two-year community college, and two inner-city Black high schools—is proposing to implement the Harvard Core Calculus Course. MARCC enrolls over 2,000 minority students in its Calculus I and II courses.

**Calculator-Enhanced Instruction Project by a Consortium of New Jersey and Pennsylvania Educational Institutions**

J. Lane                          Award No: DUE 9252491
Union County College             FY 1992 $ 229,734
Cranford, NJ 07016               Calculus

This two-year project addresses reform of the mathematics curriculum, particularly in calculus and precalculus, by incorporating the use of graphics and symbolic manipulation calculators to enhance teaching and learning following the Clemson model. The consortium will include seven two-year colleges, one private four-year college and three high schools. Faculty will attend intensive, hands-on weekend workshops on the T1-81 and HP-48 calculators in early fall 1992. Participants will convene again for four additional weekend workshops scheduled throughout 1993 and 1994.

**A Reformed Calculus Program Based on Mathematica and Project CALC**

William H. Barker               Award No: DUE 9249589
Bowdoin College                 FY 1990 $ 35,000
Brunswick, ME 04011             FY 1991 $ 44,000
                                FY 1992 $ 3,878
                                Calculus

Students are learning calculus in a discovery-based laboratory course using materials developed at Duke University and adapted for use in a liberal arts college setting. The course and laboratory materials are made available for Macintosh computers and exploit the Notebook feature of the computer algebra system Mathematica.

**Calculus Reform Workshops**

Donald B. Small                  Award No: DUE 9253119
Mathematical Association of America  FY 1992 $ 66,313
Washington, DC 20001-0000        Calculus

Three Calculus Reform Workshops were held during the summer of 1992. Each five-day workshop had 26 participants and included the following: an overview of calculus reform projects, extensive participant work with specific reformed approach, experienced curriculum reformers as work-shop instructors, participant development of curriculum materials, and establishment of a support network among participants.

**Implementation and Dissemination of the Harvard Consortium Materials in Arizona, Oklahoma, and Utah**

David Lovelock                   Award No: DUE 9252521
University of Arizona            FY 1992 $ 181,365
Tucson, AZ 85721                 FY 1993 $ 297,031
                                 FY 1994 $ 321,604
                                 Calculus

Our goal is to have substantial implementation of the Harvard Calculus Consortium materials throughout Arizona, Oklahoma, Utah, and the surrounding regions by the end of the project period. This will be done in two complementary steps. First, over the full three years, the coalition (Arizona State University, Brigham Young University, Northern Arizona University, Oklahoma State University, and the University of Arizona) will implement reform calculus. Second, during the last two years, we will expand our coalition to include satellites, which are other two- and four-year institutions and high schools from the region eager to implement reform calculus. The effort will include preliminary discussions, series of workshops, in-site visits, and electronic networking.

**A Video on Using Supercalculators in Curriculum Reform**

John W. Kenelly                  Award No: DUE 9252524
Clemson University               FY 1992 $ 60,000
Clemson, SC 29634                Calculus

A video on using supercalculators as one approach to curriculum reform has been designed, developed, tested, and evaluated. The video illustrates the changes that take place in mathematics classrooms when personal supercalculators are used regularly for instruction, homework, and tests. The completed video was mailed, free-of-charge, to each of the nation's 2,5000 mathematics departments in fall 1992.

**Workshops for Dissemination of Calculus Reform Projects**

A. Wayne Roberts                 Award No: DUE 9252529
Macalester College               FY 1992 $ 163,515
Saint Paul, MN 55105             FY 1993 $ 168,520
                                 Calculus

We plan to conduct 16 one-week 24-person workshops about calculus reform projects, eight in the summer of 1993, eight more in the summer of 1994, at sites across the country. Workshops will be run by leaders in the calculus reform movement who will describe their projects and how they overcame obstacles (need for new resources, skepticism of client disciplines, colleague resistance) that confront any curricular reform. Participants will be asked to consider their own situation and to formulate a plan for action in their home institution.

### A New Calculus Program at the University of Michigan

| Morton Brown | FY 1993 $ 324,081 |
|---|---|
| University of Michigan - Ann Arbor | FY 1992 $ 175,000 |
| Ann Arbor, MI 481091220 | FY 1993 $ 125,000 |
| | FY 1994 $ 100,000 |
| | Calculus |

The University of Michigan is redesigning its first-year calculus curriculum in content, delivery, and style. The principal features of the new program are cooperative learning, greater use of technology, increase in geometric visualization, new syllabus emphasizing problem solving, and quantitative reasoning. Starting from some pilots sections, all calculus sections will be reformed by fall 1994. There will be an extensive training program for course instructors, including a handbook and supplementary materials for use along with the Harvard Consortium text.

### Calculus and the Bridge To Calculus

| Mulloy Robertson | Award No: DUE 9252501 |
|---|---|
| Volunteer State Community College | FY 1992 $ 49,994 |
| Gallatin, TN 37066 | Calculus |

Volunteer State Community College is collaborating with area high schools and vocational schools in analyzing all math courses and their outcomes as a basis for recommending curricular changes that could result in higher-level math skills for high school graduates. Two workshops are planned The first will identify the problems, and analyze and discuss them. A goal is a 25% reduction of students who need developmental math from these specific high schools.

### Fully Renewed Calculus at Three Large Universities

| John Unbehaun | Award No: DUE 9253958 |
|---|---|
| University of Wisconsin- La Crosse | FY 1992 S 59,009 |
| La Crosse, WI 54601 | FY 1993 $ 30,178 |
| | FY 1994 $ 26,113 |
| | Calculus |

The University of Wisconsin - La Crosse is revising, testing, and refining the renewed calculus materials developed at the University of Iowa. The University of Wisconsin - La Crosse will fully implement in all calculus classes these calculus materials. We believe that this reformed calculus approach will better prepare our students to pursue careers in science and technology by being proficient at basic calculus skills, while also learning computing with Mathematica and applying the ideas of calculus in a variety of large and small open-ended projects.

### Fully Renewed Calculus at Three Large Universities

| Gurcharan S. Gill | Award No: DUE 9253959 |
|---|---|
| Brigham Young University | FY 1992 $ 18,235 |
| Provo, UT 84602 | FY 1993 $ 32,225 |
| | FY 1994 $ 33,540 |
| | Calculus |

This collaborative project is being implemented at the University of Iowa, University of Wisconsin - La Crosse, and Brigham Young University. Renewed calculus materials developed through the NSF-supported project at the University of Iowa are being revised, tested, and refined at these three institutions. The project involves extensive training of faculty and teaching assistants to use these materials and approaches. In addition, ideas are introduced through large open-ended projects on a variety of scientific and mathematical problems.

# PROJECT ABSTRACTS: FY 1993 AWARDS

## The Western Pennsylvania Calculus Technology Consortium

Frank Beatrous         Award No: DUE 9352874
University of Pittsburgh         FY 1993 $144,443
Pittsburgh, PA 15260         Mathematics

A technology-enhanced calculus courses is being established at two campuses of the University of Pittsburgh. The approach adopted in the University courses is based on the Calculus & Mathematica (C&M) project developed at the University of Illinois and at Ohio State University. The CALC-TECH project will facilitate large scale implementation of the C&M project on the two University campuses. This will require workshops for faculty and graduate teaching assistants, development of materials, and evaluation

## Maricopa Mathematics Consortium Project

Alfredo G. de los Santos      Award No: DUE 9352897
Maricopa County CC District     FY 1993 $100,000
Tempe, AZ 85281         Mathematics

The Maricopa Mathematics Consortium (M C)—composed of the Maricopa County Community College District, Arizona State University, and four public school districts in Maricopa County—is instituting a two-year project that is resulting in significant systemic change in the teaching/ learning process in precalculus mathematics. The participating institutions, which serve 235,000 students, have a long history of cooperation and joint development. This collaborative effort includes restructuring the curriculum, developing materials to reflect the changes to be made, and using technology and new pedagogies. Faculty/staff development is being provided during all phases to build faculty support and confidence in these new approaches to teaching mathematics (776 faculty). During the evaluation process, they are examining outcomes at key benchmark points in the areas of faculty development, student learning, and developmental processes.

## Development Site for Complex, Technology-Based Problems in Calculus

Brian J. Winkel         Award No: DUE 9352849
Rose-Hulman Institute of Technology    FY 1993 $100,000
Terre Haute, IN 47803         Mathematics

Faculty and students at Rose-Hulman Institute of Technology in cooperation with several high school teachers are developing and disseminating complex problems in calculus. These technology-based problems are interdisciplinary in nature and integrate concepts from calculus, engineering, and physics. The project will produce Mathematica resources to accompany problems, and writing up guidelines for using problems in high school and college calculus settings. Suggestions for nonMathematica users are being offered in each situation as well as guides on how to use the materials.

## Mathematica Laboratory Projects Projects for Calculus and Applied Mathematics

William Barker         Award No: DUE 9352868
Bowdoin College         FY 1993 $ 77,871
Brunswick, ME 04011         Mathematics

Building on Bowdoin College's three years of experience as the Mathematica laboratory development site for Project CALC, this project is to design and construct Mathematica computer laboratory notebooks for multivariable calculus and applied mathematics. The primary focus of the development efforts is on applications from disciplines which are new to the calculus and applied mathematics curriculum. Dissemination will involve a workshop as well as electronic networks. These notebooks are distinguished by their complete development of topics under study, their focus on models from other disciplines, and their use of graphics routines to develop geometric understanding.

## Interactive Modules For Courses Following Calculus

Lawrence C. Moore         Award No: DUE 9352889
Duke University         FY 1993 $ 181,827
Durham, NC 27708         FY I994 $ 180,860
                          Mathematics

The authors are developing interactive laboratory and classroom materials to support follow-on courses for students completing reformed calculus course: linear algebra, differential equations, and applied mathematical analysis. A seven-member development team, working at five different institutions over a two-year period, will develop approximately 70 interactive text modules. Each module will be developed in one of the systems Mathcad 3.1, Maple V, or Mathematica, and translated by student assistants to each of the other two systems. This project will build on and reinforce the acquired concepts and shared experiences of students moving on from reformed calculus courses. These students know how to work in a lab, how to work with partners, and how to work on their own. Moreover, they have had introductions to many of the important ideas that will be developed further in the follow-on courses. Such students will find conventional courses less than satisfactory as preparation for their scientific or engineering careers in the 21st century.

### Full-Scale Calculus Revitalization and Evaluation Project

Joe A. Marlin                          Award No: DUE 9352845
North Carolina State University        FY 1993 $ 99,878
Raleigh, NC 27695-8208                 Mathematics

An experience-based, application-driven, two-year calculus sequence involving 50 faculty members, 50 graduate teaching assistants, and over 4,000 undergraduate mathematics, science, and engineering majors each semester is being developed. The three-year plan includes revitalizing undergraduate calculus curricula and instructional techniques through curriculum enhancement, experiential computer laboratories, and intensive faculty development. The project demonstrates innovative adaptations of small-scale models previously developed by others, in addition to a new set of curriculum enhancements emphasizing applications in other disciplines, which may be adopted or further adapted at institutions of any size.

### Implementation of the Harvard Core Calculus at Stony Brook

Anthony V. Phillips                    Award No: DUE 9352843
SUNY-Stony Brook                       FY 1993 $ 155,864
Stony Brook, NY 11794                  Mathematics

The Harvard Consortium calculus is being implemented throughout the calculus courses, from beginning precalculus/calculus through multivariate calculus, in the context of a thorough revision of our first-year and remedial classes. This involves adapting the Harvard curriculum to meet the needs of a diverse student population, who fall into three categories: the 1600 students a year who take the first term of the Precalculus/Calculus sequence, for whom the precalculus end of the curriculum needs to be expanded; the 900 who take either the slow-stream or the mainstream calculus, who can use the curriculum essentially in its present form; and a smaller number of well-prepared students, who took some calculus in high school, and would benefit from enrichment material to supplement the more routine sections of the curriculum.

### A Three Semester Integrated Calculus/Physics Sequence

Wesley Ostertag                        Award No: DUE 9352941
SUNY Dutchess County College           FY 1993 $ 95,010
Poughkeepsie, NY 12601                 Mathematics

The investigators are writing laboratory materials for and team-teaching an integrated three-semester sequence of courses in calculus and physics. The core of the project is the creation of a series of mini-labs which develop analytical topics in tandem with physical principles using data-gathering equipment connected to personal computers. The investigators are considering the long-standing pair of problems in introductory science education: applications meant to motivate the calculus are often developed poorly and/or out of context by the calculus instructor; and mathematical tools needed in the physics course are often used by the physics instructor before they have been adequately developed in the calculus course.

### Mid-Atlantic Regional Calculus Consortium

Joshua Leslie                          Award No: DUE 9352865
Howard University                      FY 1993 $ 100,000
Washington, DC 20059                   FY 1994 $100,000
                                       FY 1995 $100,000
                                       Mathematics

This project is being funded as part of the Alliance for Minority Participation project, a consortium effort initially involving Howard University, Hampton University, Morgan State University, and the University of the District of Columbia. The goal of HAMP is to double within five years the number of individuals from minority groups underrepresented in science, engineering, and mathematics who obtain bachelor of science degrees in these fields. The calculus activities, centering around the Harvard Consortium reform text, will complement the strong student support which will be provided through HAMP in achieving this goal. Plans are also outlined for developing a precalculus course consistent with the Harvard Consortium materials. Student work in groups modeled after the PDP approach.

### Calculus, Linear Algebra, and ODE's in a Real and Complex World

Franklin Wattenberg                    Award No: DUE 9352828
University of Massachusetts-Amherst    FY 1993 $ 79,871
Amherst, MA 01003                      Mathematics

Being continued is the development, dissemination, and assessment of an integrated two-year sequence to replace five courses—Calculus I, II, and III, Linear Algebra, and Ordinary Differential Equations. The motivating idea driving the project is that these subjects should be taught in the context of real and engaging problems. By studying these subjects in context, students are better able to understand them and to use them outside the mathematics classroom. Emphasized are very substantial and realistic problems that require the application of a combination of mathematical techniques and that are open-ended, often without clean, simple solutions. Computers and writing are essential parts of the approach. There is a synergy between writing and meaningful problems that engages students on a high intellectual plane requiring them to understand and articulate the mathematics, the applications, and the connections between them.

### Large-Scale Calculus Revision at Penn

Dennis M. DeTurck      Award No: DUE 9352824
University of Pennsylvania      FY 1993 $100,000
Philadelphia, PA 19104      Mathematics

A full-scale revision of the entire Calculus program is being undertaken making extensive use of student projects drawn from scientific and business disciplines. The three main thrusts of the revision are (1) substantial reconstruction of the first- and second-year Calculus syllabi; (2) emphasis upon collaborative learning techniques; and (3) use of computation as a vehicle to encourage collaborative learning. The emphasis is on designing and implementing means of presenting new approaches to Calculus effectively on a large scale. Materials are being adapted from existing Calculus reform efforts. In addition, materials and support are being developed for faculty and students getting used to using computers to do mathematics and to incorporating problems from other disciplines.

### The Washington Center Calculus Dissemination Project

Robert S. Cole      Award No: DUE 9352900
Evergreen State College      FY 1993 $252,506
Olympia, WA 98505      Mathematics

The Washington Center for Improving the Quality of Undergraduate Education, is augmenting and extending for another two years the work of its Washington Center Calculus Project. The goals for the project are to build sustainable institutional commitment to calculus reform efforts already initiated, to deepen understanding of assessment tools appropriate to some of the new pedagogies being used in teaching calculus, to increase the number of institutions using reform calculus curricula, and to document the dissemination model they are using for statewide calculus reform. This proposal broadens the scope of the work in which they are currently engaged, and focuses more upon sustaining and institutionalizing curricular reform, rather than initiating it, and upon developing and documenting assessment and dissemination methods appropriate to regional initiatives.

### Differential Equations: A Dynamical Systems Approach

Paul R. Blanchard      Award No: DUE 9352833
Boston University      FY 1993 $117,260
Boston, MA 02215      FY 1994 $102,241
     Mathematics

A large-scale revision of the traditional sophomore-level ordinary differential equations course is being developed which emphasizes qualitative theory with a distinct dynamical systems orientation. A discussion of difference equations will precede the introduction of differential equations, the computer will be used to analyze solutions, and a more detailed discussion of nonlinear systems will be included. A consortium of colleges and universities will assist in developing the syllabus and materials, as well as serve as initial test sites for the textbook.

### Calculus For Comprehensive Universities and Two-Year Colleges

Gregory D. Foley      Award No: DUE 9352894
Sam Houston State University      FY 1993 $ 70,000
Huntsville, TX 77341      Mathematics

The goals of this project are to design and implement a curriculum for the first year of calculus at a set of eight comprehensive universities and two-year colleges. The program adapts and synthesizes methods from successful calculus reform efforts and is crafted to meet local needs. The curriculum stresses and facilitates cooperative learning, develops visual thinking with the aid of interactive graphing technology, and uses writing to help students learn and communicate mathematics. These methods are integrated in an environment that focuses on the central ideas of calculus and provides a progression of problems and projects to challenge students while improving their confidence and study skills. A volume of student assignments will be compiled to serve as a resource for college faculty across the nation.

### Bridge Calculus Consortium Based at Harvard

Deborah Hughes-Hallett
Harvard University
Cambridge, MA 02138

Award No: DUE 9352905
FY 1994 $141,847
FY 1995 $259,810
FY 1996 $299,212
FY 1997 $299,135

The Calculus Consortium based at Harvard University, with funding from the National Science Foundation, has developed, tested, and disseminated an innovative single-variable calculus course. In this project, the effort is being expanded to include precalculus and the second year of calculus. A major component of the proposed project is dissemination. The dissemination effort for the proposed project is being modeled on the workshops, minicourses, newsletters, test sites, and networking that have proved successful in the current project.

# PROJECT ABSTRACTS: FY 1994 AWARDS

## Interactive Electronic Third Semester Calculus Laboratory Materials for Personal Computers

Thomas F. Banchoff     Award No: DUE 9450721
Brown University     FY 1994 $181,650
Providence, RI 02912     Mathematics

Third-semester calculus of two and more variables can often be taught most effectively when classroom lectures and texts are supplemented by interactive computer graphics laboratories. The laboratory materials already developed at Brown University for introductory differential geometry using a network of SUN workstations are now being redesigned, using recent electronic book technology, so that it can run on Macintosh level computers. This is making the visualization software available to a much larger range of institutions and students, in engineering, natural sciences, life sciences, and economics, as well as computer science and mathematics.

## A Computer Based Introductory Differential Equations Course

Robert L. Borrelli     Award No: DUE 9450742
Harvey Mudd College     FY 1994 $250,127
Claremont, CA 91711     Mathematics

The aim of this project is to design a syllabus, and modules to support that syllabus, which adds computing and mathematical modeling as essential components for the introductory ordinary differential equations (ODE) course. Many modules are be self-contained units whose main focus is a topic involving the construction and analysis of a mathematical model using a computer-based approach. Some modules are designed to shed new light on traditional topics found in introductory ODE courses. The material is being produced and field-tested by a Consortium of seven institutions and is being made available to the authors of the next generation of textbooks on differential equations. The overall administration of this project is at Harvey Mudd College, but the work will be done at all seven Consortium institutions: Cornell University, Harvey Mudd College (CA), Rensselaer Polytechnic Institute, St. Olaf College (MN), Washington State University, Stetson University, and West Valley College (CA). Building on the network and experience accumulated during a previous DUE grant on faculty enhancement, the Consortium is broadening the scope of its newsletter CoODEoE to publish information about the new ODE course syllabus and modules before, during and after test-runs. In addition, CoODEoE is disseminating

information about other ODE projects around the country. A textbook publisher is adding additional support for workshops, computing programming expertise, and evaluation.

## Reforming Calculus Instruction in Puerto Rico

Rafael Martinez Planell     Award No: DUE 9450758
U of Puerto Rico Mayaguez     FY 1994 $197,935
Mayaguez, PR 00709     Mathematics

Through this project, a coalition of two- and four-year colleges and universities in Puerto Rico is changing the way calculus is taught. In addition to preparing faculty to use the Core Calculus Curriculum materials, the project is promoting and assessing the use of collaborative learning techniques with the Hispanic student population. The materials produced as well as the assessment of student achievement will also be of value to universities with large Hispanic populations.

## A Comprehensive Calculus Project for Comprehensive Universities

Curtis C. McKnight     Award No: DUE 9450760
U of Oklahoma     FY 1994 $209,992
Norman, OK 73019     Mathematics

Reformed calculus materials developed by Ostebee and Zorn at St. Olaf College are being used in the teaching of first-year calculus at The University of Oklahoma. Faculty in the Department of Mathematics are extending this model of calculus reform by undertaking tasks representing adaption, assessment, faculty and graduate student development, components for pre-service teachers, and dissemination. The St. Olaf first-year calculus model is being adapted and extended to fit a large, comprehensive university setting including the range of difficulties implied by this (large lecture with recitation, GTA-taught classes, honors classes in an open admission university, etc.).

The project includes the development of a model for GTA training along with supporting materials and training videotapes, a model for faculty development, including materials, suitable for sustaining such a calculus program, and careful assessment of student performance. Reports on assessment and evaluation will be useful to other comprehensive universities seeking similar reforms. An important component addresses pre-service secondary mathematics students: adaptation of the GTA training materials for prospective mathematics teachers; training experiments involving pre-service students; participation in GTA training sessions; and observations both in university classes and in a local high school which has adopted the St. Olaf's model for calculus. Dissemination plans include packaging materials for dissemination, production of written materials, and visits and locally held workshops. At least one monograph will be prepared for publication.

### A Regional Center for Calculus Reform at Northeastern University

Terence J. Gaffney                    Award No: DUE 9450764
Northeastern University                    FY 1994 $ 299,973
Boston, MA 02115                              Mathematics

A regional center for calculus reform is being established in the Boston area centered at Northeastern University. As part of this effort, curriculum reforms are being made to the multivariable and differential equation courses, and throughout the full range of calculus and precalculus courses. The completed reforms will affect over 150 sections; over 15,000 students will enroll in these courses in the next five years. Working with Northeastern's Comprehensive Regional Center for Minorities, the project is strengthening and helping create precalculus and calculus programs in public high schools in the City of Boston, and other urban school districts with large minority enrollments. As an incentive for students and teachers to join this program, Northeastern will award credit to all high school students completing a calculus course through this program. There will also be a regional network of school teachers and college faculty committed to the adaptation and dissemination of the ideas and techniques of the reform movement, reaching thousands of area students. This network will maintain a software and materials library and sponsor seminars, workshops, and regional conferences.

### Multivariable Calculus from Graphical, Numerical, and Symbolic Points of View

Arnold M. Ostebee                    Award No: DUE 9450765
Saint Olaf College                          FY 1994 $ 150,061
Northfield, MN 55057                          Mathematics

A curriculum and materials development project focusing on multivariate calculus builds on and extends the work by the PI's on single variable calculus. Faculty at St. Olaf College are writing, field testing, and disseminating complete course materials, including a textbook, for a rethought multivariable calculus course that emphasizes the development of conceptual understanding and analytical reasoning skills. The course and its supporting materials share a pervasive mathematical theme and a consistent pedagogical strategy: to combine, compare, and move among graphical, numerical, and algebraic viewpoints on calculus. Graphical, numerical, and symbolic computing technology is used to support this emphasis. The materials use technology freely but are not bound to any specific hardware or software platform. The technology requirements are defined generically in terms of functionalities, rather than by brand name. The text assumes that students have access to technology that has at least the capabilities of a CAS. Specific applications to a particular CAS can be found in supplements and lab manuals.

### West Point Core Curriculum Conference in Mathematics

Donald Small                              Award No: DUE 9450767
US Military Academy                         FY 1994 $ 95,000
West Point, NY 10996                          Mathematics

The CUPM Recommendations for a General Mathematical Sciences Program detailed a new view of undergraduate mathematics, stating a philosophy centering on goals for student learning and outlining a needed broadening of the curriculum. Subsequent CUPM documents call for a four semester sequence of study that contains the core of mathematics normally found in three semesters of calculus and semesters of discrete mathematics, linear algebra, differential equations, and probability and statistics. This call for a broader view of mathematics dovetails well with the vision for secondary school mathematics outlined in the NCTM's Curriculum and Evaluation Standards. A workshop during summer, 1994, brings together leaders from secondary, community college, liberal arts, and research university mathematics to discuss such 7-into-4 programs and to review their implications for the variety of individuals and client disciplines they serve.

The United States Military Academy at West Point has a fully implemented model of a 7-into-4 curriculum. The activities of the workshop focus on discussing the content that should be central to each of the seven curricular areas, examining the Academy's program, and considering how content can be better integrated to better prepare the students served. The products of the conference consist of proceedings with commissioned papers and reactions, details on the West Point program, and actions and recommendations of the workshop. Panel discussions based on the workshop will be proposed for future MAA meetings.

### Gateways to Advanced Mathematical Thinking: Linear Algebra and Precalculus

| | |
|---|---|
| Al Cuoco | Award No: DUE 9450731 |
| Education Development Ctr | FY 1994 $236,029 |
| Newton, MA 02160 | FY 1995 $273,177 |
| | FY 1996 $272,977 |

This project is building on previous research and curriculum development work to develop flexible understandings of topics in precalculus, calculus, and linear algebra. Exemplary course materials will be produced using a modular approach to package the concepts and activities. Based on the broad mathematical themes, these materials will make use of constructivist pedagogies, involving cooperative learning, computer technology, and alternatives to traditional lecturing. Field testing of the curriculum will provide sites for research into the way students learn the topics and environments for teacher enhancement.

### Graphing Calculator Resource Materials for Calculus and Pre-Calculus

| | |
|---|---|
| W. Frank Ward | Award No: DUE 9450744 |
| Indian River Cmty College | FY 1994 $100,000 |
| Fort Pierce, FL 33454 | Mathematics |

Faculty in the Department of Mathematics at Indian River Community College are developing instructional materials for the use of the graphing calculator to support teaching with the Harvard Calculus Consortium materials that are appropriate for a two-year college audience. Materials are being developed for both teachers and students. The teacher supplement contains examples and instructions on how and when to use a graphing calculator effectively as an instructional tool. Less emphasis is placed on traditional lecture, and more emphasis is placed on exploration and group problem-solving. The material integrates practical applications taken from astronomy and physics and other areas that use mathematics. Students generate real data through the use of a Calculator Base Laboratory System (CBL). The device enables the collection of real-time data for analysis with the graphing calculator. The collection and analysis of real data provides the means to integrate practical laboratory experience into pre-calculus and calculus. The instructional materials include the use of the CBL system.

The student supplement is designed with an emphasis on using technology to explore mathematical concepts and write about the findings. Projects address four types of activities: (1) "get started" activities to familiarize students with using the calculators; (2) activities to help students understand concepts; (3) projects to promote deeper understanding of concepts that may be used as group projects or for advanced students; and (4) investigation projects for topics beyond what is discussed in class. This range of activities will benefit the slower student by providing activities for understanding as well as the advanced student by providing enrichment material. The conceptual approach combined with real applications, emphasis on understanding, and problem solving, benefits all students.

### Instituting Calculus Reform: A Community College State University Consortium Model

| | |
|---|---|
| William L. Lepowsky | Award No: DUE 9450735 |
| Peralta Cmty Col Dist Off | FY 1994 $119,703 |
| Oakland, CA 94606 | Mathematics |

The Peralta Community College District, San Francisco City College, California State University Hayward, and San Francisco State University are forming a consortium to jointly revitalize instruction in calculus by adapting the Harvard Calculus Curriculum. Faculty from each institution are jointly developing the curriculum which will incorporate instructional methods and strategies that have proven to be effective in increasing the success rate of calculus students. The results are being published in a Calculus Reform Handbook. The respective mathematics departments are working to institutionalize the developed reform curriculum in all calculus classes at the four institutions. In addition, the core group is developing a preparation program for mathematics faculty who have not participated in the curriculum development effort.

### Calculus, Concepts, Computers and Cooperative Learning: Assessment and Evaluation in Terms of Dissemination Goals

| | |
|---|---|
| Edward L. Dubinsky | Award No: DUE 9450750 |
| Purdue Univ Research Fdn | FY 1994 $50,000 |
| West Lafayette, IN 47907 | FY 1995 $168,000 |
| | Mathematics |

This project is engaging in assessment, evaluation, and research into how students learn mathematical concepts to determine effects of the calculus reform curriculum and dissemination activities from the previously supported Calculus project at Purdue. Assessment activities will be closely related to research into how students learn mathematical concepts and will also include studies of the effect of the role of the teacher when changed from instructor to facilitator and the use of group learning. Current dissemination approach will include the preparation of an

Instructor's Resource Manual, workshops, and the continued growth and development of a network of implementers. A model will also be developed for evaluation of student attitudes, performance, and conceptual understanding in comparison with students who took calculus in standard models.

### Assessing the Workshop Calculus with Review Project

Nancy H. Baxter                     Award No: DUE 9450746
Dickinson College                      FY 1994 $ 100,505
Carlisle, PA 17013                              Mathematics

Workshop Calculus with Review I and II is a unique sequence of courses which integrates a review of fundamental concepts needed to explore ideas in calculus with the study of concepts normally encountered in Calculus I. Materials have been developed with support from the NSF ILI program and the US Department of Education FIPSE program. The goals of this two-year assessment project are to: (1) identify a list of "community norms", that is, a list of key concepts students who take an integrated pre-calculus/calculus sequence are expected to know; (2) design and implement internal and external assessment tools to evaluate student learning gains and attitudes, with a particular emphasis on the impact of technology; and (3) track student retention rates, continuation rates, and performance in subsequent mathematics classes and classes in other disciplines that have a calculus prerequisite. Materials are being beta-tested in a variety of settings and developers are assisting colleagues in adapting the materials for use at their institutions. Based on outcomes of the assessments undertaken by this project, the Workshop Calculus with I and II instructional materials will be refined and published for widespread dissemination. The results of the project will be published, and the assessment tools will be distributed.

Workshop Calculus with Review I and II is a integration of pre-calculus concepts with Calculus I material and is designed for students who are not prepared to enter Calculus I. The sequence provides a gateway into the traditional calculus sequence. In the Workshop environment, lectures are replaced by interactive teaching, where students learn by doing and by reflecting on what they have done. In the workshop environment, lectures are replaced by interactive teaching in which no formal distinction is made between classroom and laboratory work. In class, students explore mathematical ideas on their own and discover their own approaches to solving problems, using technology as appropriate and by working collaboratively on tasks in their Student Activity Guides.